Cambridge International GCSE

Maths

The key to success in the Cambridge International GCSE Maths exams is practice. Luckily for you, this brilliant CGP book is full of exam-style questions to work through so you're fully prepared.

It covers both the Core and Extended exams, and it's all up to date for the latest specification — nice!

There are even fully worked answers in the back, so if you drop any marks, it's easy to find out exactly where you went wrong.

Exam Practice Workbook

Contents

✓ Use the tick boxes to check off the topics you've completed.

How to Use This Book..1
Exam Tips..2

Section One — Number

Order of Operations..3
Ordering by Size and Negative Numbers...4
Special Types of Number...5
Multiples and Factors..6
Prime Numbers and Prime Factors..7
LCM and HCF...8
Fractions...9
Fractions, Decimals and Percentages...11
Percentages...12
Compound Growth and Decay..15
Ratios..17
Proportion...19
Rounding Numbers...21
Bounds..22
Standard Form...23
Sets and Venn Diagrams..25

Section Two — Algebra

Algebra Basics...28
Powers..29
Expanding Brackets...30
Factorising...31
Manipulating Surds...32
Solving Equations...33
Formulas...35
Algebraic Fractions...37
Factorising Quadratics..38
The Quadratic Formula...40
Completing the Square...41
Simultaneous Equations..42
Solving Equations Using Graphs...44
Inequalities...46
Graphical Inequalities...47
Sequences...48
Algebraic Proportion...51

Section Three — Graphs, Functions and Calculus

Coordinates...52
Straight-Line Graphs...53
Quadratic Graphs..56
Harder Graphs..58
Real-Life Graphs...60
Travel Graphs...61
Functions..63
Differentiation..64

Section Four — Geometry and Measures

Geometry 66
Bearings and Scale Drawings 68
Polygons and Symmetry 70
Circle Geometry 72
Congruence and Similarity 75
The Four Transformations 76
More Enlargements 78
Unit Conversions 79
Time 80
Speed, Density and Pressure 81
Perimeter and Area 83
Triangle Constructions 85
3D Shapes — Surface Area and Nets 86
3D Shapes — Volume 88

Section Five — Pythagoras, Trigonometry and Vectors

Pythagoras' Theorem 90
Trigonometry — Sin, Cos and Tan 91
The Sine and Cosine Rules 93
Trig Graphs 95
3D Pythagoras 96
3D Trigonometry 97
Vectors 98

Section Six — Probability and Statistics

Probability Basics 100
Finding Probabilities and Expected Frequency 101
The AND/OR Rules 103
Tree Diagrams 104
Probability from Venn Diagrams 106
Relative Frequency 107
Organising Data 108
Mean, Median, Mode and Range 109
Frequency Tables 110
Grouped Frequency Tables 111
Simple Charts 112
Pie Charts 114
Scatter Diagrams 115
Interpreting and Comparing Data 116
Histograms 118
Cumulative Frequency 120

Mixed Questions 122

Answers 136

Formulas in the Exams 164

Published by CGP

Editors:
Liam Dyer, Jake McGuffie, Caley Simpson, Julie Wakeling, Hannah Wilkie

Contributor:
Alastair Duncombe

With thanks to Maddie Wright for the proofreading.
With thanks to Jan Greenway for the copyright research.

ISBN: 978 1 83774 121 2

Clipart from Corel®
Printed by W&G Baird Ltd, Antrim.

Based on the classic CGP style created by Richard Parsons.

Text, design, layout and original illustrations © Coordination Group Publications Ltd. (CGP) 2023
All rights reserved.

Photocopying this book is not permitted, even if you hold a licence.
Extra copies are available from CGP • www.cgpbooks.co.uk

How to Use This Book

- Hold the book <u>upright</u>, approximately <u>50 cm</u> from your face, ensuring that the text looks like <u>this</u>, not ꜱᴉɥʇ.
  Alternatively, place the book on a <u>horizontal</u> surface (e.g. a table or desk) and sit adjacent to the book, at a distance which doesn't make the text too small to read.
- In case of emergency, press the two halves of the book together <u>firmly</u> in order to close.
- Before attempting to use this book, familiarise yourself with the following <u>safety information</u>:

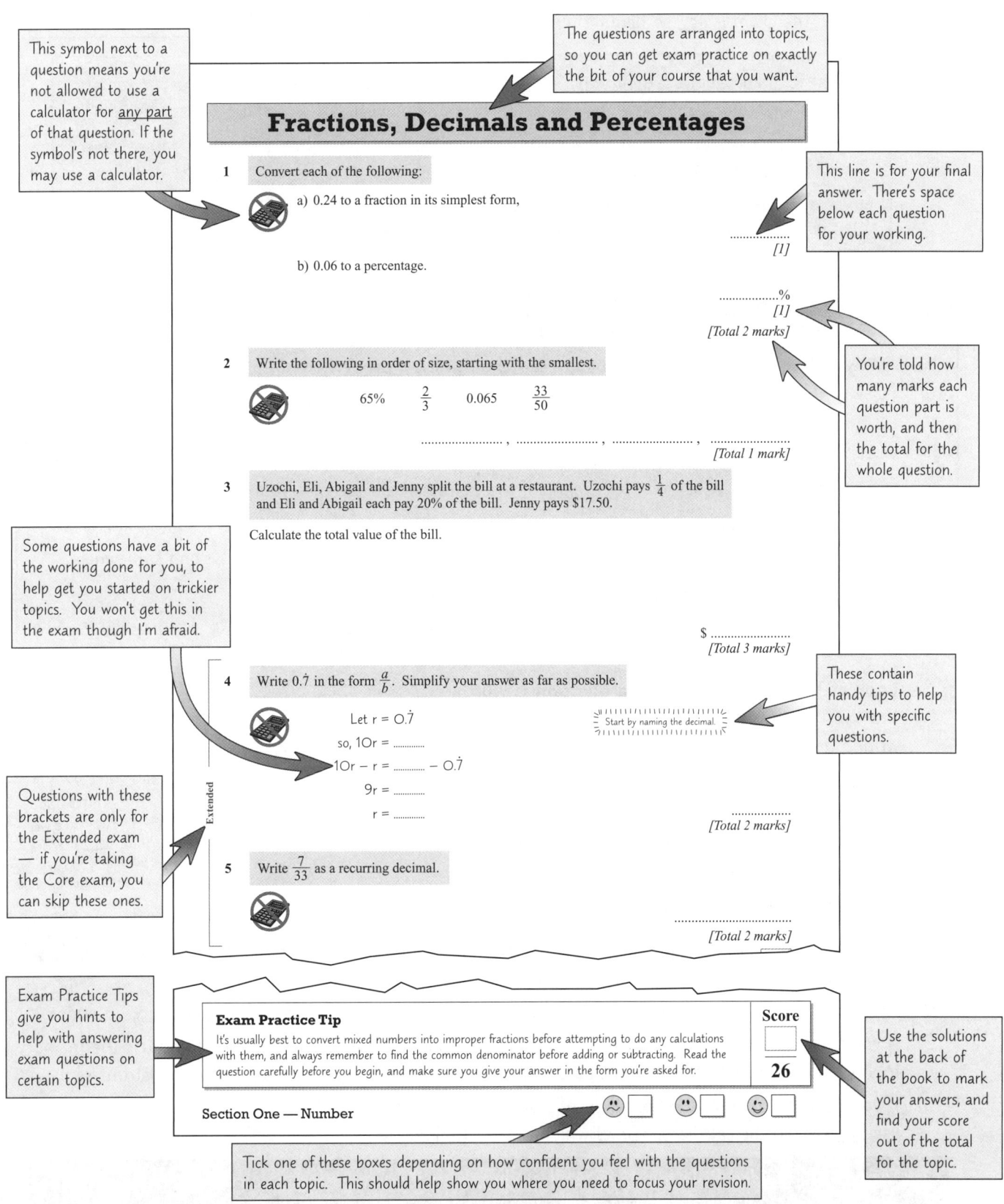

Exam Tips

Exam Stuff

1) You'll do <u>two</u> exam papers — one <u>non-calculator paper</u> and one <u>calculator paper</u>.

2) Both papers have a <u>mixture</u> of short, <u>unstructured</u> questions and long, <u>structured</u> questions.

3) <u>Timing</u> in the exam is <u>key</u>, so here's a quick guide...

	Core	Extended
Non-calculator Paper	<u>Paper 1</u> 80 marks, 1.5 hours	<u>Paper 2</u> 100 marks, 2 hours
Calculator Paper	<u>Paper 3</u> 80 marks, 1.5 hours	<u>Paper 4</u> 100 marks, 2 hours

- Aim to spend about a <u>minute per mark</u> working on each question (i.e. 2 marks = 2 mins). Don't spend ages and ages on a question that's only worth a few marks.
- If you have any time left at the end of the exam, use it to <u>check</u> back through your answers and make sure you haven't made any silly mistakes.
- If you're totally, hopelessly stuck on a question, just <u>leave it</u> and <u>move on</u> to the next one. You can always <u>go back</u> to it at the end if you've got enough time.

There are a Few Golden Rules

1) **Always, always, always make sure you <u>read the question properly</u>.**
   For example, if the question asks you to give your answer in metres, <u>don't</u> give it in centimetres.

2) **Show <u>each step</u> in your <u>working</u>.**
   You're less likely to make a mistake if you write things out in stages. And even if your final answer's wrong, you'll probably pick up <u>some marks</u> if the examiner can see that your <u>method</u> is right.

3) **Check that your answer is <u>sensible</u>.**
   Worked out an angle of 450° or 0.045° in a triangle? You've probably gone wrong somewhere...

4) **Make sure you give your answer to the right <u>degree of accuracy</u>.**
   The question might ask you to round to a certain number of <u>significant figures</u> or <u>decimal places</u>. So make sure you do just that, otherwise you'll almost certainly lose marks.

5) **Look at the number of <u>marks</u> a question is worth.**
   If a question's worth 2 or more marks, you probably won't get them all for just writing down the final answer — you're going to have to <u>show your working</u>.

6) **Write your answers as <u>clearly</u> as you can.**
   If the examiner can't read your answer you won't get any marks, even if it's right.

Obeying these Golden Rules will help you get as many marks as you can in the exam — but they're no use if you haven't learnt the stuff in the first place. So make sure you revise well and do <u>as many</u> practice questions as you can.

Using Your Calculator

1) Your calculator can make questions a lot easier for you, but only if you <u>know how to use it</u>. Make sure you know what the different buttons do and how to use them.

2) Remember to check your calculator is in <u>degrees mode</u>. This is important for <u>trigonometry</u> questions.

3) If you're working out a <u>big calculation</u> on your calculator, it's best to do it in <u>stages</u> and use the <u>memory</u> to store the answers to the different parts. If you try and do it all in one go, it's too easy to mess it up.

4) If you do want to do a question all in one go on your calculator, use <u>brackets</u> so the calculator knows which bits to do first.

REMEMBER: <u>Golden Rule number 2</u> still applies, even if you're using a calculator — you should still write down <u>all</u> the steps you are doing so the examiner can see the method you're using.

Section One — Number

Order of Operations

1 Work out the following.

a) $4 \times 11 + 14 \div 2 \times 3 - 2$

....
[1]

b) $(3 \times 20 - 15) - (21 + 35 \div 7)$

....
[1]
[Total 2 marks]

2 Serena has written out the calculation below.

Add one pair of brackets to Serena's calculation to make it correct.

$3 - 2 \times 7 + 9 \div 3 + 11 = -6$

[Total 1 mark]

3 Use your calculator to work out the value of $\dfrac{(13 \times 6 + 5)^2 + \sqrt{(6 \times 4) + (50 \times 20)}}{3^2}$.

....
[Total 1 mark]

4 Use your calculator to work out the value of $\dfrac{197.8}{\sqrt{0.01} + 0.23}$.

Write down all the figures on your calculator display in your answer.

....
[Total 1 mark]

5 x and y are integers and $0 < x < y$.

Write down two sets of values for x and y such that $3 = \sqrt{x^2 - 2y}$.

> Start off by getting rid of the square root, then try values of x > 0.

$x = $ .... , $y = $ ....

or $x = $ .... , $y = $ ....
[Total 2 marks]

Exam Practice Tip

Knowing what order to do your operations in makes maths much easier. If you're struggling to remember the order, try writing BODMAS (Brackets, Other, Division, Multiplication, Addition, Subtraction) on the first page of your exam paper. That way, you can always look back at it when you need to do a calculation.

Score

7

Ordering by Size and Negative Numbers

1 Choose four different numbers from the list below to complete the following statements.

   3   8   2   6   9   5

7 + ... < 3 + ... − 2   −4 + ... + 2 ≥ .... + 3

[Total 2 marks]

2 Put the numbers below in order from lowest to highest.

   0.75   −0.23   −0.61   0.35   1.06   −1.12

.... , ... , ... , ... , ... , ...

[Total 1 mark]

3 Find the reciprocal of −5. Give your answer as a decimal.

....

[Total 1 mark]

4 Put the numbers below in order from smallest to biggest.

   0.35   0.035   0.53   0.335   0.355   0.0355   0.503

.... , ... , ... , ... , ... , ... , ...

[Total 1 mark]

5 Tiago measures the outdoor temperature at his house at 11 am every Saturday for 10 weeks.

Work out the difference between the temperatures in week 2 and week 9.

Week	1	2	3	4	5	6	7	8	9	10
Temperature (°C)	−2	−4	0	−1	4	7	5	10	14	16

.... °C

[Total 1 mark]

6 Noor has three number cards. She is going to use the numbers to make a new number.

   −4   3   5

She can use each number once and the operations +, −, ×, ÷ and brackets.
Find the highest number she can make.

....

[Total 2 marks]

Score:

8

Section One — Number

Special Types of Number

1 Find the value of $0.54^6 + \sqrt[3]{2.87}$. Give your answer to 3 significant figures.

....
[Total 1 mark]

2 Look at the numbers below.

18   5.5   1.75   –22 $\sqrt{3}$

1.333... 2.562^2 5π $\sqrt{16}$

a) Write down the integers.

....
[1]

b) Write down the irrational numbers.

....
[1]
[Total 2 marks]

3 State whether the following expression is rational or irrational.

 $\dfrac{\sqrt{6}}{4\sqrt{(10 - 2 \times 2)}}$

The expression is ....
[Total 2 marks]

4 Find:

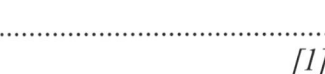
Start off by finding the roots of the numbers in the question.

a) the square number between 270 and 300,

....
[1]

b) the cube number between 320 and 360.

....
[1]
[Total 2 marks]

Score:
7

Section One — Number

Multiples and Factors

1 Look at the list of numbers below.

    80   66   64   72   62   74

a) Write down a number from the list that is a multiple of 12.

....
[1]

b) Write down a number from the list that is a factor of 128.

....
[1]
[Total 2 marks]

2 Write down:

a) all the factors of 28,

....
[1]

b) all the multiples of 8 which appear in the list below.

    55 56 57 58 59 60 61 62 63 64 65

....
[1]
[Total 2 marks]

3 Eric says, "even square numbers always have more factors than odd square numbers." Find examples to show that Eric is wrong.

[Total 2 marks]

4 A number, x, is a common factor of 252 and 420, and a common multiple of 6 and 7. Given that $x > 50$, find the value of x.

$x = $ ....
[Total 3 marks]

Score:

9

Section One — Number

Prime Numbers and Prime Factors

1 Look at the list of numbers below.

                1   7   11   12   15   21

a) Write down a number from the list which is a prime number.

....
[1]

b) Write down two numbers from the list whose sum is a prime number.

.... and ....
[1]

c) Write down a number from the list which is a prime factor of 84.

....
[1]
[Total 3 marks]

2 Jack says, "there are no prime numbers between 100 and 110."
Is he correct? Give evidence for your answer.

[Total 2 marks]

3 Jayanti thinks of a two-digit prime number.
The sum of its digits is one more than a square number.

Write down one number Jayanti could be thinking of.

....
[Total 2 marks]

4 Write 72 as a product of its prime factors.

Make sure your answer only uses prime numbers. Multiply them all together to check you get the number you started with.

....
[Total 2 marks]

Score:

LCM and HCF

1 Write down the highest common factor (HCF) of 12 and 32.

....
[Total 1 mark]

2 $P = 3^7 \times 11^2$ and $Q = 3^4 \times 7^3 \times 11$.

Write as the product of prime factors:

a) the lowest common multiple (LCM) of P and Q,

For the LCM, multiply together the highest power of each factor that appears in P and Q. For the HCF, multiply together the factors that appear in both P and Q.

....
[1]

b) the highest common factor (HCF) of P and Q.

....
[1]
[Total 2 marks]

3 Find the lowest common multiple (LCM) of 15 and 24.

....
[Total 2 marks]

4 Philippa is making jam.

She needs to buy mini jam jars which come in packs of 16 and labels which come in packs of 36. She doesn't want to have any items left over.

Find the smallest number of packs of each item she can buy.

.... packs of jars and .... packs of labels
[Total 3 marks]

Score:

8

Fractions

1   Write down a fraction that is equivalent to $\frac{16}{21}$.

....
[Total 1 mark]

2   Find 40 as a fraction of 15.

   Give your answer as a mixed number in its simplest form.

....
[Total 2 marks]

3   Work out the following.

   Give your answers as mixed numbers.

> Make sure each fraction has the same denominator before you add or subtract.

a) $3\frac{1}{2} + 2\frac{3}{5}$

....
[3]

b) $3\frac{3}{4} - 2\frac{1}{3}$

....
[3]
[Total 6 marks]

4   Work out the following.

   Give your answers as mixed numbers in their simplest form.

a) $1\frac{2}{3} \times \frac{9}{10}$

....
[3]

b) $3\frac{1}{2} \div 1\frac{2}{5}$

....
[3]
[Total 6 marks]

5 Shapes X, Y and Z are shown below.

$\frac{2}{5}$ of shape X is shaded and $\frac{6}{7}$ of shape Y is shaded.

Calculate the fraction of shape Z that is shaded.

....
[Total 3 marks]

6 Musa and his 4 friends ate $\frac{5}{6}$ of a pizza each. Pizzas cost $12.50 each, or 2 for $22.

 Calculate the minimum amount that Musa and his friends will have spent on pizzas.

$ ....
[Total 3 marks]

7 A factory buys 25 tonnes of flour. $17\frac{1}{2}$ tonnes of the flour is used to make scones.

$\frac{1}{5}$ of the flour used for scones is used to make cheese scones.

What fraction of the total amount of flour is used to make cheese scones?

....
[Total 2 marks]

8 *ABC* is an equilateral triangle. It has been divided into smaller equilateral triangles as shown below.

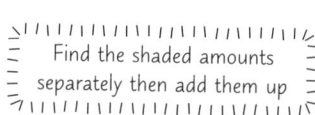

 Calculate the fraction of triangle *ABC* that is shaded.

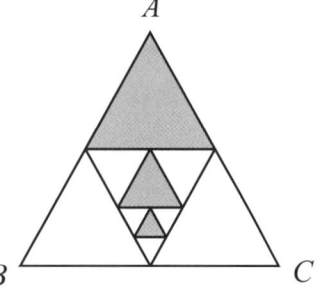

....
[Total 3 marks]

Exam Practice Tip

It's usually best to convert mixed numbers into improper fractions before attempting to do any calculations with them, and always remember to find the common denominator before adding or subtracting. Read the question carefully before you begin, and make sure you give your answer in the form you're asked for.

Score

26

Fractions, Decimals and Percentages

1 Convert each of the following:

a) 0.24 to a fraction in its simplest form,

....
[1]

b) 0.06 to a percentage.

.................%
[1]
[Total 2 marks]

2 Write the following in order of size, starting with the smallest.

$\quad$ 65% $\quad\quad \dfrac{2}{3} \quad\quad$ 0.065 $\quad\quad \dfrac{33}{50}$

.... , ... , ... , ...
[Total 1 mark]

3 Uzochi, Eli, Abigail and Jenny split the bill at a restaurant. Uzochi pays $\dfrac{1}{4}$ of the bill and Eli and Abigail each pay 20% of the bill. Jenny pays $17.50.

Calculate the total value of the bill.

$ ....
[Total 3 marks]

4 Write $0.\dot{7}$ in the form $\dfrac{a}{b}$. Simplify your answer as far as possible.

$\quad\quad$ Let r = 0.$\dot{7}$

$\quad\quad$ so, 10r = ...

$\quad\quad$ 10r − r = ... − 0.$\dot{7}$

$\quad\quad\quad\quad$ 9r = ...

$\quad\quad\quad\quad\quad$ r = ...

Start by naming the decimal.

....
[Total 2 marks]

5 Write $\dfrac{7}{33}$ as a recurring decimal.

....
[Total 2 marks]

Score: ☐
10

Percentages

1 The ratio of grapes to cherries in a fruit salad is 2 : 5.
Circle the correct statement below.

There are 50% more cherries than grapes.   There are 80% more cherries than grapes.

There are 20% as many grapes as cherries.   There are 40% as many grapes as cherries.

[Total 1 mark]

2 Alessandra is completing a 240-piece jigsaw.
72 of the jigsaw pieces show monkeys.

Work out the percentage of jigsaw pieces that show monkeys.

.............................%
[Total 1 mark]

3 A school library contains 278 fiction books and 197 non-fiction books.

200 books are borrowed from the library.
Now 44% of the books in the library are fiction books.
How many fiction books were borrowed from the library?

Hint: start by working out the total number of books remaining in the library.

....
[Total 3 marks]

4 A computer costs $927 in a sale.
Next week, the cost of the computer will increase by 20%.

Calculate the cost of the computer next week.

$ ....
[Total 2 marks]

Section One — Number

5 Sofia and two friends are booking festival tickets online using their credit cards.
Tickets cost $180 each, plus an additional charge of $5.40 per credit card transaction.

a) Calculate the percentage increase in the cost of buying one ticket if it's bought using a credit card.

.... %
[2]

b) Calculate the percentage saving if Sofia and her friends buy three tickets in one transaction rather than three separate transactions. Give your answer to 2 d.p.

.... %
[3]
[Total 5 marks]

6 Jane invests $326 in an account with a simple interest rate of 1.5%.

How much money will be in Jane's account after 7 years?

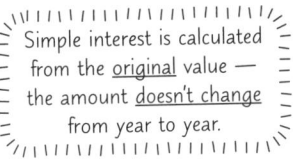
Simple interest is calculated from the original value — the amount doesn't change from year to year.

$ ....
[Total 3 marks]

7 In the triangular prism below, the base and vertical height of the triangular face are x cm and the length of the prism is y cm.

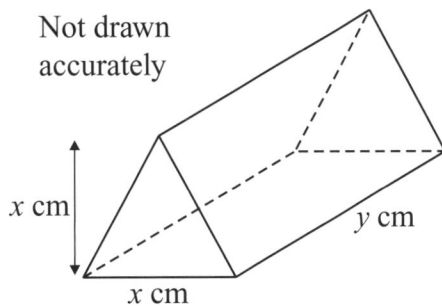
Not drawn accurately

Work out the percentage increase in the volume of the prism when x is increased by 15%.

.... %
[Total 3 marks]

Section One — Number

8 After an 8% pay rise Mr Mokhtar's salary was $28 728.

Work out Mr Mokhtar's salary before the increase.

$ ....
[Total 3 marks]

9 Ian makes and sells lobster pots. He sells them for $32 per pot, which is a 60% profit on the cost of making a pot.

a) Calculate how much it costs Ian to make each lobster pot.

$ ....
[3]

Ian wants to increase his profits, so he increases the price of each lobster pot to $37.60.

b) Calculate the percentage profit Ian will now make on each lobster pot.

........................%
[3]
[Total 6 marks]

Exam Practice Tip

As you can see, questions about percentages are often based on real-life situations, such as taxes or prices, so it can often be helpful to check if your answers sound realistic. For example, if you've worked out that a chocolate bar has increased in price by $1 000 000, it might be worth checking your numbers again.

Score

27

Compound Growth and Decay

1 The balance of a savings account, B, is given by the formula $B = 5000 \times 1.02^t$ where t is the number of years since the account was opened.

a) Work out the balance of the account when it was first opened.

$ ....
[1]

b) Calculate the amount of money in the account after 7 years.
Give your answer to the nearest cent.

$ ....
[2]
[Total 3 marks]

2 Rich inherits $10 000 and wants to invest it. His bank is offering him two accounts.

Compound Collectors Account	**Simple Savers Account**
5.5% compound interest per year	6.2% simple interest per year

Determine which account will give him the most money after 5 years.

....
[Total 4 marks]

3 Mrs Khan puts $2500 into a high interest savings account.
Compound interest is added to the account at the end of each year.
After 2 years Mrs Khan's account contains $2704.

Calculate the interest rate on Mrs Khan's account.

.... %
[Total 3 marks]

Section One — Number

4 Joshua buys a 4-year-old laptop for $482.18.
He works out that the value of his laptop has depreciated by 29% each year.

Work out how much Joshua's laptop was worth 4 years ago.

$ ....
[Total 3 marks]

5 The population of fish in a lake is estimated to decrease by 8% every year.

a) Estimate how many fish will be left after 15 years if the initial population is 2000.

....
[2]

b) Estimate how many years it will take for the population of fish to be less than $\frac{3}{4}$ of the initial population.

$\frac{3}{4}$ of the initial population = ...

2000 × .... = ...

.... × ....^{2} = ...

.... × ....^{3} = ...

.... × ....^{4} = ...

.... years
[2]
[Total 4 marks]

Ratios

1 Achara is making some green paint to paint her kitchen wall.
She makes it by mixing together $3\frac{3}{4}$ tins of yellow paint and $1\frac{1}{2}$ tins of blue paint.
The tins are all the same size.

a) Write the ratio of yellow paint to blue paint in its simplest form.

....
[2]

b) Work out how much of each paint Achara will need to make 2800 ml of green paint.

yellow paint .... ml

blue paint .... ml
[2]
[Total 4 marks]

2 Chocolate milkshake is made by mixing milk and ice cream in the ratio $2:9$.

a) Calculate the amount of milk used as a fraction of the ice cream used.

....
[1]

b) Calculate the amount of milkshake made if 801 ml of ice cream is used.

.... ml
[2]
[Total 3 marks]

3 Eve is making a bird house. To make the walls, she takes a plank of wood and cuts it into four pieces in the ratio $5:6:6:7$. The longest wall is 9 cm longer than the shortest wall.

Calculate the length of the original plank of wood.

.... cm
[Total 2 marks]

4 Brown sauce can be bought in three different sizes.
The price of each is shown on the right.
Which size of bottle is the best value for money?

.... ml
[Total 3 marks]

5 Simone, Ariana, Nasir and Catrin shared $660.
Nasir got four times as much money as Catrin, Simone got twice as much money as Nasir, and Ariana got a quarter as much money as Simone.

 How much money did Simone get?

$ ....
[Total 3 marks]

6 In a zoo, the ratio of wolves : bears = 7 : 4.
The ratio of wolves : giraffes = 11 : 8.

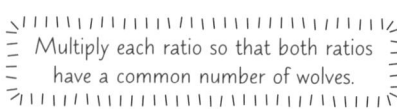
Multiply each ratio so that both ratios have a common number of wolves.

Work out the ratio of bears : giraffes.

bears : giraffes = ... : ....
[Total 2 marks]

Score:

17

Section One — Number

Proportion

1 Joanna gets paid the same hourly rate whenever she works.
In the first week of July, Joanna worked for 28 hours and got paid $231.
In each of the next 3 weeks of July, she worked for 25 hours.

Work out how much Joanna will get paid in total for the 4 weeks she worked in July.

$ ....
[Total 2 marks]

2 Ishmael is making some T-shirts. It takes 5 m^2 of cotton to make 8 T-shirts.
Cotton costs $2.75 per square metre.

Work out how much it will cost Ishmael to buy enough cotton to make 85 T-shirts.
Give your answer to the nearest cent.

1 T-shirt will take: .... ÷ .... = ... m^2 of cotton

85 T-shirts will take: .... × .... = ... m^2 of cotton

.... m^2 of cotton costs .... × .... = $................

$ ....
[Total 3 marks]

3 It takes 12 people 3 hours to harvest the crop from a field.

Estimate how long it would take 4 people to harvest the crop.

.... hours
[Total 2 marks]

4 Yusef and Sophie are knitting some socks. Sophie needs to knit three times as many socks as Yusef but she can knit them twice as quickly.

Yusef takes 26.5 hours to knit his socks.
Work out how long it would take Sophie to knit her socks.

.... hours
[Total 2 marks]

Section One — Number

5 Cat bakes 18 sponge cakes to sell.
The recipe on the right will make 5 sponge cakes.

Ingredients
275 g flour (plain)
275 g butter
220 g sugar
5 eggs (medium)

a) Calculate how much of each ingredient she uses.

Flour: .... g

Butter: .... g

Sugar: .... g

Eggs: ....
[3]

The total cost of the ingredients for 18 sponge cakes is $25.30.
She cuts each cake into 10 slices and sells the slices for 50c each.

b) Work out the profit that Cat will make if she sells every slice.

$ ....
[3]
[Total 6 marks]

6 A ship has enough food to cater for 250 people for 6 days.

a) For how many days can the ship cater for 300 people?

.... days
[2]

b) Work out how many more people it can cater for on a 2-day cruise than on a 6-day cruise.

.... people
[3]
[Total 5 marks]

Exam Practice Tip
There are two key tips for proportion questions. Firstly, don't get confused between direct and inverse. Directly proportional quantities increase together, inversely proportional quantities go opposite ways. Secondly, find the value for one then scale it up if it's direct proportion, or down if it's inverse proportion.

Score

20

Rounding Numbers

1 The man in this picture is 176 cm tall.

Use this information to estimate the height of the penguin.

Think about what would be a reasonable degree of accuracy for your estimate.

.... cm
[Total 2 marks]

2 The distance between two stars is 428.6237 light years.

Write this distance correct to 2 significant figures.

.... light years
[Total 1 mark]

3 A stall sells paperback and hardback books. Paperback books cost $4.95 and hardback books cost $11. One Saturday, the stall sells 28 paperback and 19 hardback books.

a) Find an estimate for the amount of money the stall took that day.
Show all your working.

$ ....
[2]

b) The actual amount the stall took was $347.60.
Do you think your estimate was sensible? Explain your answer.

....

....
[1]
[Total 3 marks]

4 Work out an estimate for $\sqrt{\dfrac{2321}{19.673 \times 3.81}}$

Show all of your working.

Round the numbers to 1 significant figure before doing the calculation.

....
[Total 2 marks]

Score: 8

Section One — Number

Bounds

1 Joseph is weighing himself. His scales give his weight to the nearest kilogram.

According to his scales, Joseph is 57 kg.
What are the minimum and maximum weights that he could be?

Minimum weight: .... kg

Maximum weight: .... kg
[Total 2 marks]

2 Given that $x = 2.2$ correct to 1 decimal place, complete the statement below about the value of x.

.... ≤ x < ....
[Total 2 marks]

3 Eric is comparing the volume of two buckets. He measures the volume of each bucket to the nearest 0.1 litres and finds that bucket A has a volume of 8.3 litres and bucket B has a volume of 13.7 litres.

Calculate the lower bound of the difference, in litres, between the volumes of bucket A and bucket B.

.... litres
[Total 2 marks]

4 Rounded to 1 decimal place, a triangle has a height of 3.2 cm and an area of 5.2 cm². Calculate the upper bound for the base length of the triangle, giving your answer to 2 d.p.

.... cm
[Total 3 marks]

Score: ☐
9

Standard Form

1 Write 907 200 000 in standard form.

....
[Total 1 mark]

2 A plane flies 1.27×10^4 km.

Write down the distance in words.

.... kilometres
[Total 2 marks]

3 Light travels at approximately 1.86×10^5 miles per second.
The distance from the Earth to the Sun is approximately 9.3×10^7 miles.

Time = Distance ÷ Speed

Calculate the time it will take light to travel this distance.
Give your answer in standard form.

.... seconds
[Total 2 marks]

4 The table on the right shows the masses of four different particles.

a) Which particle is the heaviest?

....
[1]

Particle	Mass (g)
Particle A	2.1×10^{-7}
Particle B	8.6×10^{-8}
Particle C	1.4×10^{-6}
Particle D	3.2×10^{-7}

b) Write the mass of particle C as an ordinary number.

.... g
[1]

c) What is the difference in mass between particle A and particle D?
Give your answer in standard form.

.... g
[1]
[Total 3 marks]

Section One — Number

5 A patient has been prescribed a dose of 4×10^{-4} grams of a certain drug to be given daily.

a) The tablets that the hospital stocks each contain 8×10^{-5} grams of the drug.
Work out how many tablets the patient should be given each day.

Show your working.

.... tablets
[3]

b) The doctor increases the patient's daily dose of the drug by 6×10^{-5} grams.
Work out the patient's new daily dose of the drug.

Show your working and give your answer in standard form.

You need matching powers to be able to add two numbers together in standard form.

.... grams per day
[3]
[Total 6 marks]

6 The distance from the Earth to the Sun is approximately 1.5×10^8 km.
The distance from Neptune to the Sun is approximately 4.5×10^9 km.

Calculate the ratio of the Earth-Sun distance to the Neptune-Sun distance.
Give your answer in the form $1:n$.

....
[Total 2 marks]

7 Express $\dfrac{3^2}{2^{122} \times 5^{120}}$ in standard form.

$$\dfrac{3^2}{2^{122} \times 5^{120}} = \dfrac{........}{2^{......}(2^{......} \times 5^{120})}$$

$$= \dfrac{........}{........ \times 10^{......}}$$

$$= \dfrac{........}{........} \times \dfrac{1}{10^{......}}$$

$$= .... \times 10^{......}$$

....
[Total 2 marks]

Exam Practice Tip
Numbers in standard form can look scary because they're either really huge or really tiny. But don't worry — you can treat them just like any other number. Make sure you know how to type standard form numbers into your calculator before the exam, and how to interpret a standard form answer from the display.

Score: 18

Sets and Venn Diagrams

1 Write a statement to interpret the following:

a) $\xi = \{1, 2, 3, 4, 5, 6, 7\}$

....
[1]

b) $n(A \cup B) = 15$

....
[1]

c) $J \subseteq (K \cap L)$

....
[1]

d) $Z = \{\text{Square numbers between 5 and 8}\} = \varnothing$

....
[1]

[Total 4 marks]

2 If $\xi = \{3, 4, 5, 6, 7, 8, 9, 10, 11, 12, 13, 14, 15, 16\}$, $A = \{\text{Factors of 36}\}$ and $B = \{\text{Square numbers}\}$, write down the elements of $A \cap B$.

....
[Total 1 mark]

3 The Venn diagram below represents 60 elements.
39 elements are in set Y. 45 elements are in set Z.
21 elements are in set Z but not in set Y.

 Complete the diagram.

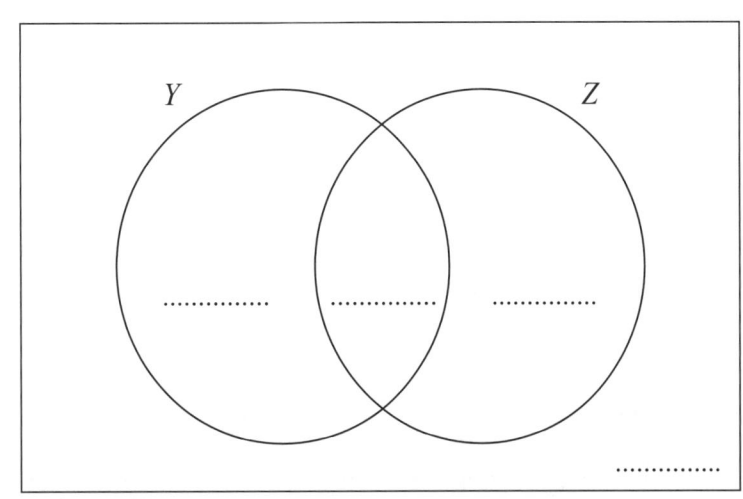

[Total 2 marks]

4 The Venn diagram below shows the number of elements in sets A and B.

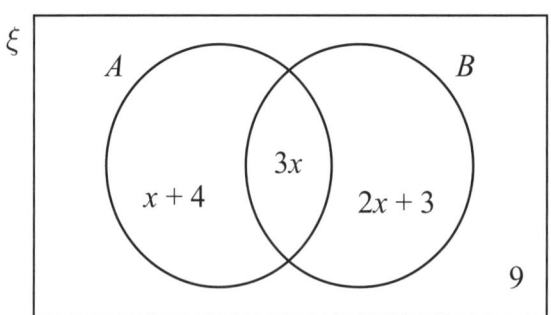

Given that $n(\xi) = 40$, find the value of x.

$x = $ ....
[Total 2 marks]

5 The Venn diagram below represents the crowd at a soccer match.

$\xi = \{\text{fans at the match}\}$
$R = \{\text{fans wearing a red shirt}\}$
$W = \{\text{fans wearing a white scarf}\}$

Given that 400 fans attended the match, $n(R) = 300$, $n(W) = 310$ and $n(R \cup W)' = 40$, complete the Venn diagram.

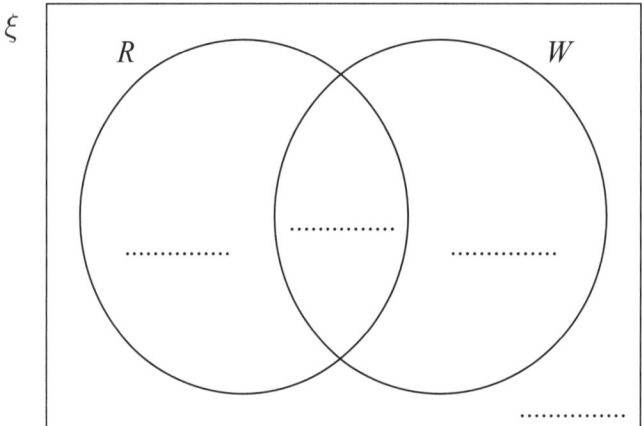

[Total 2 marks]

6 $\xi = \{\text{Integers between 5 and 50}\}$, $C = \{\text{Multiples of 5}\}$,
$D = \{\text{Multiples of 3}\}$ and $E = \{\text{Factors of 90}\}$.

a) Write down the set $C' \cap D \cap E$.

....
[2]

b) Shannon says that all the elements of $C' \cap D \cap E$ will also be elements of $(C \cup D \cup E)'$.
Is she correct? Give evidence for your answer.

....

....
[2]
[Total 4 marks]

Section One — Number

7 The Venn diagram below has been partially labelled with the number of elements.

n(ξ) = 24, n(F) = 14, n(G) = 15

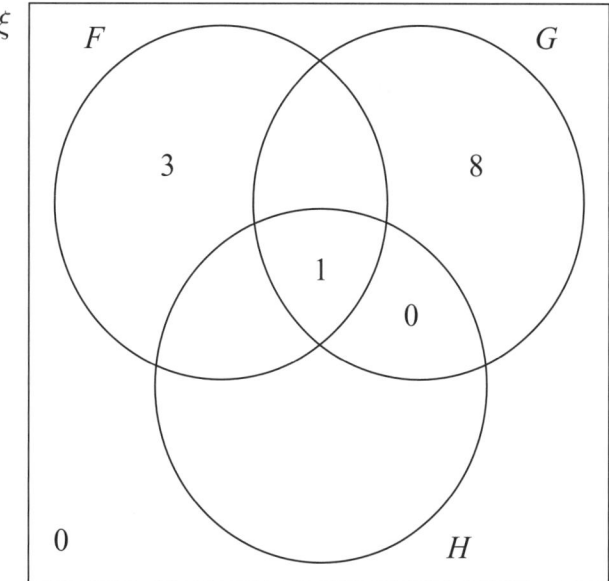

a) Shade the region of the Venn diagram described by $F \cap G \cap H'$.

[1]

b) Find the number of elements in the set $F \cap G' \cap H$.

....
[2]
[Total 3 marks]

8 ξ = {Natural numbers less than 11}, P = {x : 2 ≤ x ≤ 8},
Q = {Multiples of 3}, R = {Even numbers}.

Barry says that $(P \cap Q) \subseteq R$, while Michiko says that $(P' \cap Q) \subseteq R'$.
Determine if either of them are correct.

Start by writing down the sets you need to answer the question, then find the intersections between the sets.

[Total 4 marks]

Exam Practice Tip
There's an easy way to check the numbers in your Venn diagram, as long as you know the number of elements in the universal set. The numbers inside the rectangle should add up to the same number as the universal set. If your numbers don't match, then you may need to go and double-check your answers.

Score
22

Section One — Number

Algebra Basics

1 Circle the simplified version of $4s - 3s + 9s$.

       16s   12s   10s   −8s   11s

[Total 1 mark]

2 Simplify the following.

a) $p + p + p + p$

....
[1]

b) $7r - 2p - 4r + 6p$

....
[2]

c) $2x^2 + 8x - 4x - x^2$

....
[2]
[Total 5 marks]

3 On the diagram below, shade the area represented by $pq + 3pr$.

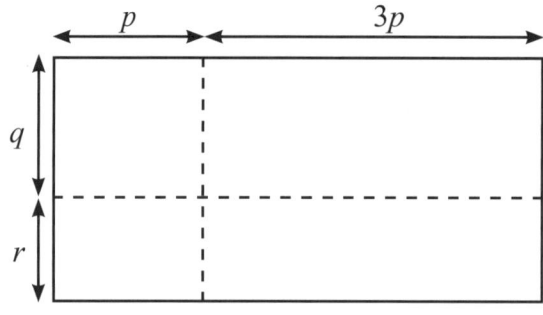

[Total 1 mark]

4 The diagram below shows a rectangle with sides that are $4x + 3$ cm and $5x - 9$ cm long.

Find an expression in terms of x for the side length of a regular hexagon with the same perimeter as the rectangle.

$5x - 9$ cm

$4x + 3$ cm

Diagram not accurately drawn

.... cm
[Total 3 marks]

Score:

10

Powers

1 Circle the correct value of 5^{-2}.

    −25   2.5 $\frac{2}{5}$ $\frac{1}{25}$ $\sqrt{5}$

[Total 1 mark]

2 For values of $y \geq 2$, write the following expressions in order from smallest to largest.

y^{-3} y^3 y^1 y^0

....

[Total 2 marks]

3 Simplify the expression $\dfrac{3^4 \times 3^7}{(3^6)^{-2}}$. Leave your answer in index form.

....

[Total 2 marks]

4 Fully simplify $(x^4 \times x^7) \div (x^3 \times x^2)^2$.

$(x^4 \times x^7) = x^{..........} = x^{..........}$

$(x^3 \times x^2) = x^{..........} = x^{..........}$, so $(x^3 \times x^2)^2 = (x^{..........})^2 = x^{..........}$

So $(x^4 \times x^7) \div (x^3 \times x^2)^2 = $ ....

....

[Total 2 marks]

5 Find the value of $16^{\frac{3}{4}}$.

....

[Total 2 marks]

6 Simplify the expressions below.

a) $\dfrac{2}{5}x^{\frac{1}{2}} \div 2x^{-2}$

....
[2]

b) $\left(\dfrac{24a^{30}}{3a^3}\right)^{\frac{1}{3}}$

....
[3]

[Total 5 marks]

Score: 14

Section Two — Algebra

Expanding Brackets

1 Expand the brackets in the following expressions.
Simplify your answers as much as possible.

   a) $5p(6 - 2p)$

....
[2]

   b) $(2t - 5)(3t + 4)$

....
[2]
[Total 4 marks]

2 a, b and c are integers such that $4(5x - 7) + 6(4 - 2x) = abx + ac$.
Suggest one set of possible values of a, b and c when $a > 0$.

$a = $ ...., $b = $ ...., $c = $ ....
[Total 3 marks]

3 Write an expression for the area of the rectangle below.
Simplify your expression as much as possible.

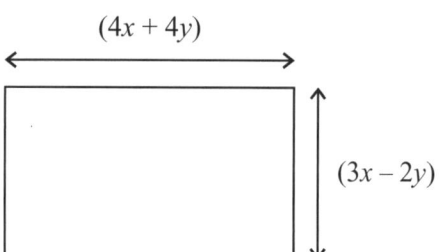

Diagram not accurately drawn

....
[Total 2 marks]

4 Expand and simplify $(5c + 6)(2c + 1)(c + 1)$.

....
[Total 3 marks]

Factorising

1 Factorise the following expressions completely.

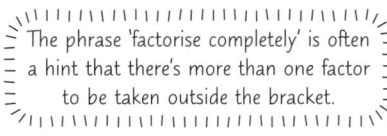
The phrase 'factorise completely' is often a hint that there's more than one factor to be taken outside the bracket.

 a) $7y - 21y^2$

....
[2]

b) $8ab^2 - 12a^2bc$

....
[2]

c) $12pr + 6qr - 30pqr$

....
[2]
[Total 6 marks]

2 Factorise the following expressions completely.

 a) $x^2 + 6xy + 9y^2$

....
[1]

b) $16p^2 - 9q^2$

....
[2]

c) $3g - 6h + 4fg - 8fh$

....
[2]
[Total 5 marks]

Extended

Exam Practice Tip
If you're worried that you haven't factorised an expression properly in the exam, you can check it by expanding the brackets back out. If you've done the factorisation correctly then you should end up with the same expression that you started with. If you don't, then go back to the start and give the factorisation another go.

Score: 11

Manipulating Surds

1 Write $(2 + \sqrt{3})(5 - \sqrt{3})$ in the form $a + b\sqrt{3}$, where a and b are integers.

....
[Total 2 marks]

2 Write $2\sqrt{50} - (\sqrt{2})^3$ in the form $a\sqrt{b}$, where a and b are integers.

....
[Total 2 marks]

3 Express $\sqrt{396} + \dfrac{22}{\sqrt{11}} - \dfrac{220}{\sqrt{44}}$ in the form $a\sqrt{11}$, where a is an integer.

....
[Total 4 marks]

4 Express $\dfrac{1 + \sqrt{7}}{3 - \sqrt{7}}$ in the form $a + b\sqrt{7}$, where a and b are integers.

> Multiply by $3 + \sqrt{7}$ to rationalise the denominator.

....
[Total 4 marks]

Exam Practice Tip

Surds can look a bit scary at first, but they're really just numbers that you can add, subtract, multiply and divide with, just like you would with any other number. Make sure you're familiar with the rules of surds — it might not be what you want to hear, but the only way to get good at these is to do lots of practice.

Score

12

Section Two — Algebra

Solving Equations

1 Solve the equations below.

a) $p - 11 = -7$

$p = $ ....
[1]

b) $3z + 2 = z + 15$

$z = $ ....
[2]
[Total 3 marks]

2 Find the solution to each of the following equations.

a) $3(a + 2) = 15$

$a = $ ....
[2]

b) $5(2b - 1) = 4(3b - 2)$

$b = $ ....
[3]
[Total 5 marks]

3 Solve the equation $(x + 2)(x - 4) = (x - 2)(x + 1)$.

Start by expanding the brackets on both sides of the equation.

$x = $ ....
[Total 4 marks]

Section Two — Algebra

4 Solve the following equations.

a) $5x^2 = 180$

$x = $ ....
[2]

b) $\dfrac{8-2x}{3} + \dfrac{2x+4}{9} = 12$

$x = $ ....
[4]
[Total 6 marks]

5 The diagram below shows an equilateral triangle.

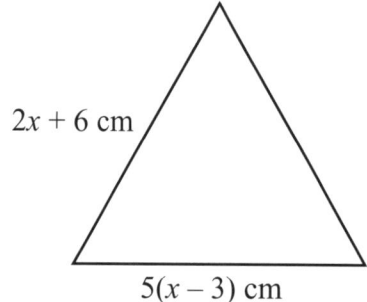

Find the length of one side of the equilateral triangle.

.... cm
[Total 4 marks]

6 Neil and Liam want to buy a new games console which costs $360. They both get weekend jobs, where they each get paid $4.50 per hour. When they've earned enough to afford the games console between them, Liam has worked 30 more hours than Neil.

How many hours did each boy work?

Neil: .... hours, Liam: .... hours
[Total 3 marks]

Score: ☐ / 25

Section Two — Algebra

Formulas

1 The formula $v = u + at$ can be used to calculate the speed of a car.

a) Rearrange the formula to make u the subject.

.... [1]

b) Rearrange the formula to make t the subject.

.... [2]

[Total 3 marks]

2 To convert kilometres into miles, Tasmin says that you divide the number of kilometres by 8 and multiply the answer by 5.

a) Write this rule as a formula.
Use k to represent the number of kilometres and m to represent the number of miles.

.... [2]

b) Use your formula to convert 110 kilometres into miles.

.... miles [2]

[Total 4 marks]

3 The formula for the distance travelled, s, by a dropped object in free fall is $s = \frac{1}{2}gt^2$, where g is the constant acceleration due to gravity and t is time taken.

Rearrange the formula to make t the subject.

.... [Total 3 marks]

Section Two — Algebra

4 Nancy, Chetna and Norman are baking cakes for a cake stall. Chetna bakes twice as many cakes as Nancy and Norman bakes 12 more cakes than Chetna. They bake 72 cakes in total.

How many cakes does each person bake?

Nancy: .... , Chetna: .... , Norman: ....
[Total 4 marks]

5 The relationship between *a*, *b* and *y* is given by the formula $a + y = \dfrac{b - y}{a}$.

   a) Rearrange this formula to make *y* the subject.

....
[4]

b) Find the value of *y* when *a* = 3 and *b* = 6.

y = ...
[1]
[Total 5 marks]

6 The following formula has *x* as the subject: $x = \sqrt{\dfrac{n-1}{2n-3}}$

a) Rearrange the formula to make *n* the subject.

....
[5]

b) Find the value of *n* if *x* = 0.2.

....
[1]
[Total 6 marks]

Exam Practice Tip
Some of the formulas you're given in the exam may look frightening, but you can rearrange them in the exact same way as you would solve a normal equation. It's also worth taking the time to carefully check each step when you're rearranging. You don't want to make a mistake that could cause you problems later on.

Score

25

Algebraic Fractions

1 Simplify the algebraic fractions below as much as possible.

a) $\dfrac{x^2 - 2x}{x^2 - 5x + 6}$

....
[3]

b) $\dfrac{4x^2 + 10x - 6}{16x^2 - 4}$

....
[3]
[Total 6 marks]

2 Combine these algebraic fractions.

$\dfrac{x+4}{2x+1} + \dfrac{2x-2}{x-2}$

....
[Total 4 marks]

3 Simplify the calculation below as much as possible.

$\dfrac{2a-8}{a^2-9} \div \dfrac{a^2-2a-8}{a^2+5a+6} \times (2a^2-a-15)$

....
[Total 5 marks]

Score:
15

Factorising Quadratics

1 Factorise the expression $x^2 + 9x + 18$ completely.

Work out which pair of numbers could be added together to give 9 and multiplied together to give 18.

....
[Total 2 marks]

2 Factorise the expression $y^2 - 4y - 5$ completely.

....
[Total 2 marks]

3 Factorise the expression $x^3 + 4x^2 - 32x$ completely.

....
[Total 3 marks]

4 Solve the equation $x^2 + 4x - 12 = 0$ by factorising.

$x = $ .... or $x = $ ....
[Total 3 marks]

5 Solve the equation $2x^2 + 13x - 7 = 0$ by factorising.

$x = $ .... or $x = $ ....
[Total 3 marks]

6 Rearrange and solve $3x = \dfrac{15 - 9x}{2x}$ by factorising.

$x = $ .... or $x = $ ....
[Total 4 marks]

7 The product of two consecutive positive even numbers is 288.
Find the larger of the two numbers by forming and solving an equation.

> Start off by forming an equation
> and then factorise to get two values.

....
[Total 4 marks]

The Quadratic Formula

1 Solve the equation $2x^2 - 7x + 2 = 0$ using the quadratic formula.

Show your working and give your answers correct to 2 decimal places.

a = ..., b = ... and c = ...

$$x = \frac{-\text{......} \pm \sqrt{\text{......}^2 - 4 \times \text{......} \times \text{......}}}{2 \times \text{......}} = \frac{\text{......} \pm \sqrt{\text{......}}}{\text{......}}$$

x = ... or x = ...

x = ... or x = ...
[Total 3 marks]

2 Solve the equation $3x^2 - 2x - 4 = 0$ using the quadratic formula.

Show your working and give your answers correct to 2 decimal places.

x = ... or x = ...
[Total 3 marks]

3 Solve the equation $6x^2 + 4x - 3 = 8$ using the quadratic formula.

Show your working and give your answers correct to 2 decimal places.

x = ... or x = ...
[Total 3 marks]

4 The area of the rectangle on the right is 30 cm². Find the length of the longer side of the rectangle.

Show your working and give your answer correct to 2 decimal places.

Form an equation and solve using the quadratic formula.

The rectangle has sides $(x+3)$ cm and $(3x+3)$ cm.

.... cm
[Total 5 marks]

Exam Practice Tip

Thankfully, you'll be given the quadratic formula in your exam — you won't need to remember it (phew). The quadratic formula looks tricky, but all you need to do is substitute the numbers in the correct places — just make sure your quadratic equation is in the correct form ($ax^2 + bx + c = 0$) before you do that.

Score 14

Completing the Square

1 Given that $x^2 + ax + b = (x + 2)^2 - 9$, work out the values of a and b.

$a =$ .... and $b =$ ....
[Total 2 marks]

2 The expression $x^2 - 10x - 5$ can be written in the form $(x + p)^2 + q$.

a) Find the values of p and q.

$p =$ .... and $q =$ ....
[2]

b) Use your answer to solve the equation $x^2 - 10x - 5 = 0$.
Leave your answer in surd form.

$x =$ .... or $x =$ ....
[2]
[Total 4 marks]

3 A curve has the equation $y = 2x^2 + 6x + 7$.

a) Write the expression $2x^2 + 6x + 7$ in the form $a(x + s)^2 + t$.

....
[2]

b) Use your answer to find the minimum point on the curve with equation $y = 2x^2 + 6x + 7$.

(...................., ...)
[2]
[Total 4 marks]

Score:
10

Simultaneous Equations

1 Solve this pair of simultaneous equations.

$3x + 4y = 26$
$2x + 2y = 14$

$2x + 2y = 14 \xrightarrow{\times 2} 4x + \text{............} = \text{............}$
$\phantom{2x + 2y = 14 \xrightarrow{\times 2}} - 3x + 4y = 26$
$\phantom{2x + 2y = 14 \xrightarrow{\times 2}............} x = \text{............}$

$\text{............} + 4y = 26$
$4y = 26 - \text{............} = \text{............}$
$y = \text{............}$

$x = \text{............} \quad y = \text{............}$
[Total 3 marks]

2 Solve this pair of simultaneous equations.

$x + 3y = 11$
$3x + y = 9$

$x = \text{............} \quad y = \text{............}$
[Total 3 marks]

3 A sweet shop sells bags of pick 'n' mix. A bag that contains 4 chocolate frogs and 3 sugar mice costs $3.69. A bag that contains 6 chocolate frogs and 2 sugar mice costs $3.96.

Work out how much a bag that contains 2 chocolate frogs and 5 sugar mice would cost.
Show your working.

$ \text{...................}$
[Total 6 marks]

4 Solve the following pair of simultaneous equations.

$x^2 + y = 4$
$y = 4x - 1$

$x = $ .... , $y = $ ....
and $x = $ .... , $y = $ ....
[Total 5 marks]

5 Solve the following pair of simultaneous equations.

$2x^2 - 4y = -14 - 12x$
$x + 2y = -5$

$x = $ .... , $y = $ ....
and $x = $ .... , $y = $ ....
[Total 5 marks]

6 The lines $y = x^2 + 3x - 1$ and $y = 2x + 5$ intersect at two points. Find the length of the line connecting the two points.

Give your answer correct to two decimal places.

Use Pythagoras' theorem to find the distance between the two points.

....
[Total 7 marks]

Exam Practice Tip

When you're solving simultaneous equations in the exam, it's always a good idea to check your answers at the end. Just substitute your values for x and y back into the original equations and see if they add up as they should. If they don't then you must have gone wrong somewhere, so go back and check your working.

Score

29

Section Two — Algebra

Solving Equations Using Graphs

1 The diagram below shows graphs of $2y - x = 5$ and $4y + 3x = 25$.

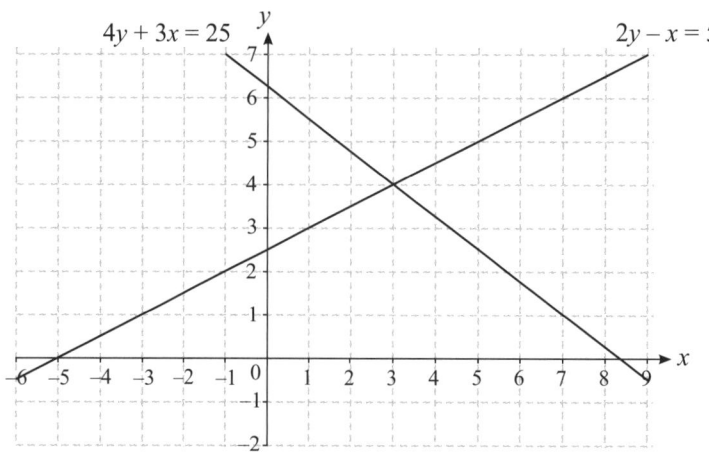

Use the diagram to solve these simultaneous equations:
$2y - x = 5$
$4y + 3x = 25$

$x = $ .... $y = $ ....
[Total 1 mark]

2 The graphs of the equations $y = \dfrac{6}{x}$ and $y = 2x - 1$ are shown below.

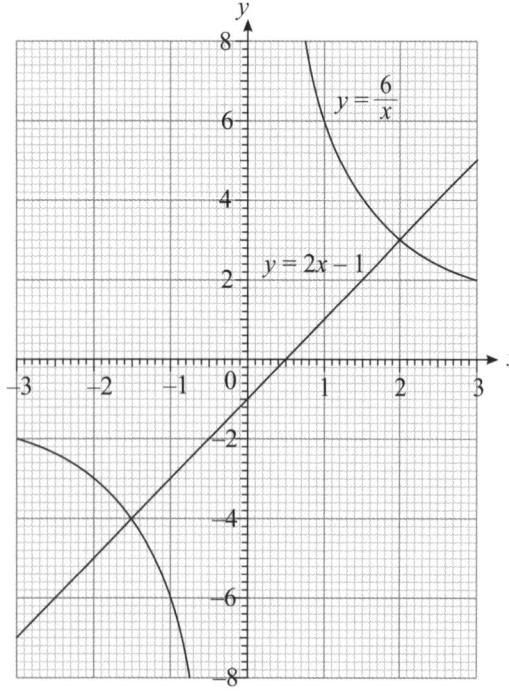

Using the graphs, write down the solutions to the simultaneous equations
$y = \dfrac{6}{x}$ and $y = 2x - 1$.

$x = $ .... , $y = $ ....

and $x = $ .... , $y = $ ....
[Total 2 marks]

Section Two — Algebra

3 The diagram below shows the lines $y = x + 1$ and $y = 4 - 2x$.

a) Use the diagram to solve $x + 1 = 4 - 2x$

x = ...
[1]

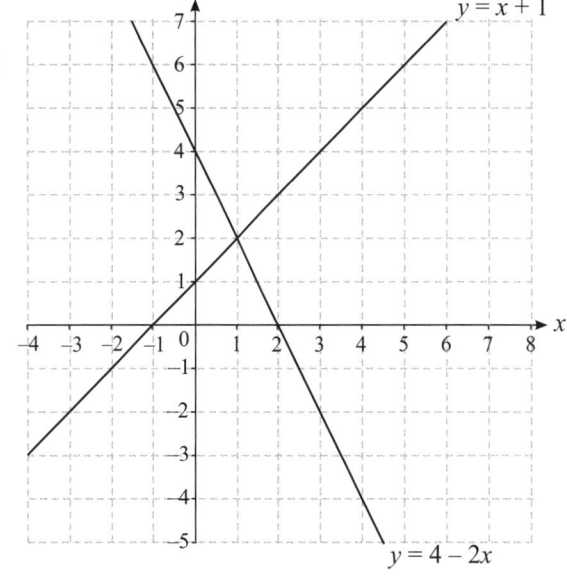

b) By drawing another straight line,
solve these simultaneous equations:

$y = x + 1$

$3y = x + 9$

x = ... y = ...
[3]

[Total 4 marks]

4 The graph of the curve $y = x^2 + 2x - 5$ is shown below.
By drawing a suitable line on the graph, find the solutions of $x^2 + x = 6$.

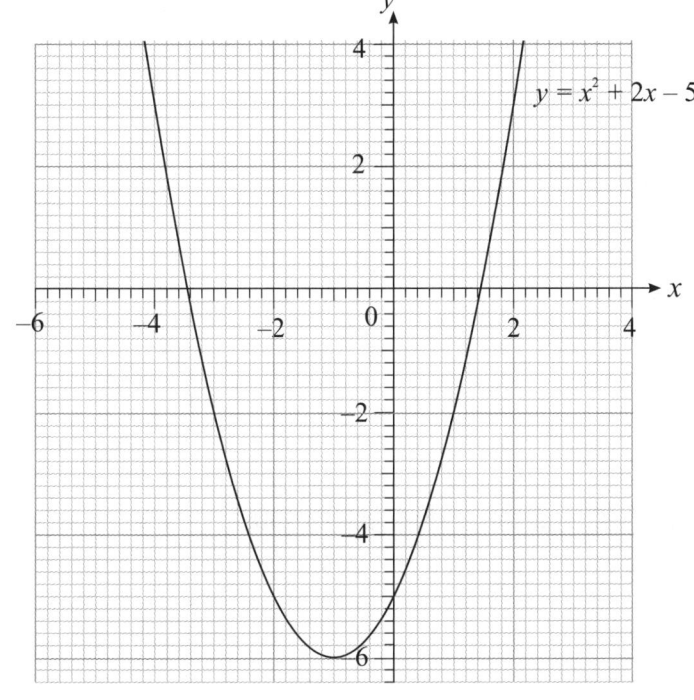

x = ... or x = ...
[Total 4 marks]

Score:
11

Inequalities

1 Write down the inequality shown on the number line below.

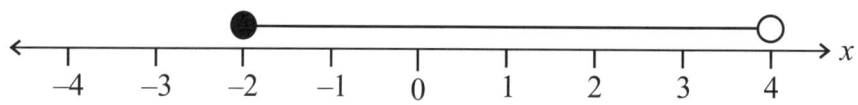

....
[Total 1 mark]

2 p and q are integers. $p \leq 45$ and $q > 25$.

What is the largest possible value of $p - q$?

....
[Total 2 marks]

3 Solve the following inequalities.

 a) $4q - 5 < 23$

....
[2]

b) $4r - 2 \geq 6r + 5$

....
[2]
[Total 4 marks]

4 Find the integer values that satisfy both of the following inequalities:

$5n - 3 \leq 17$ and $2n + 6 > 8$
Give your answer using set notation.

....
[Total 3 marks]

5 Find the largest three consecutive even numbers that sum to less than 1000.

> Start by writing an equation
> where the answer is < 1000.

....
[Total 3 marks]

Score:

13

Section Two — Algebra

Graphical Inequalities

1 Look at the grid below.

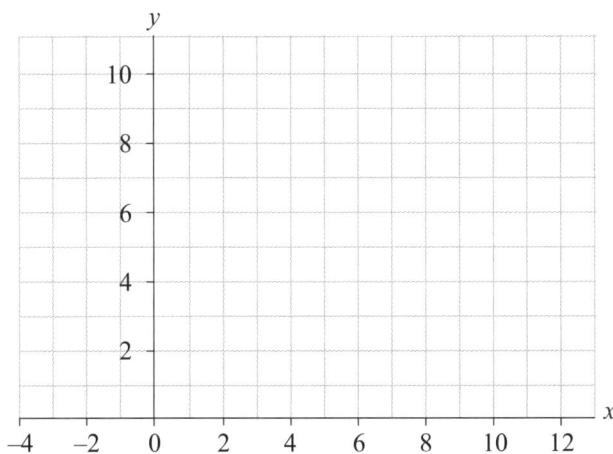

a) Use the grid to draw the graphs of $2x + y = 10$ and $y = x + 2$.

[2]

b) Draw and label, using the letter S, the area represented by the inequalities $x \geq 1$, $2x + y \leq 10$, $y \geq x + 2$.

[2]

[Total 4 marks]

2 Look at the grid on the right.

On the grid, draw the region that represents these inequalities:

$x \leq 5$

$y \geq -2$

$y - x \leq 1$

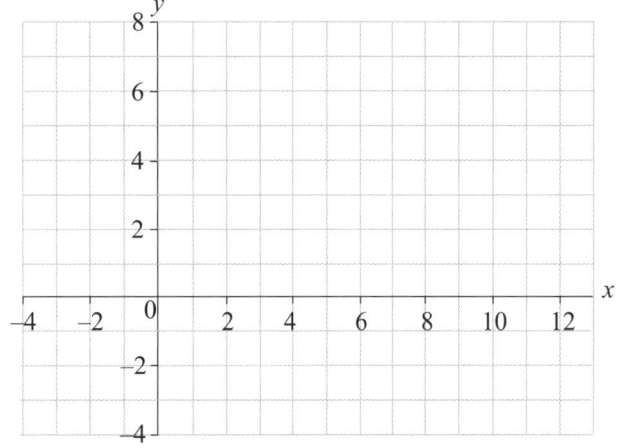

[4]
[Total 4 marks]

3 Look at the grid on the right.

The shaded region R is bounded by the lines $y = 2$, $y = x$ and $x + y = 8$.

Write down three inequalities which define R.

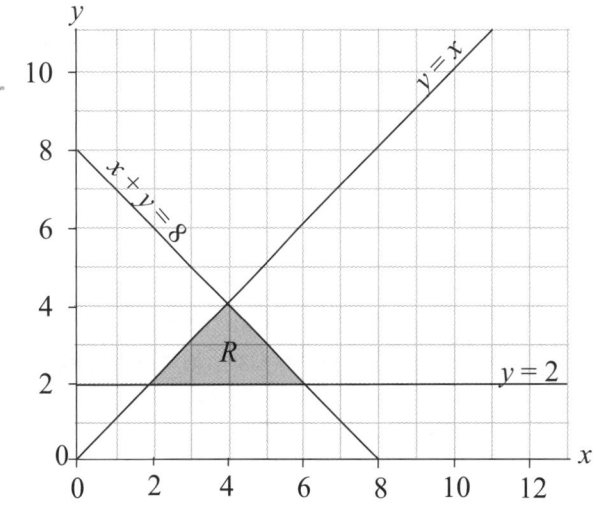

....

....

....

[Total 3 marks]

Exam Practice Tip

You need to pay close attention to whether the symbol is just < or > or whether it's ≤ or ≥. If it's < or >, draw a dashed line on the graph. If it's ≤ or ≥ you need to use a solid line. You should test your inequalities too — pick a point with coordinates that satisfies the inequality to check whether it is in the correct region.

Score: 11

Sequences

1 Here are the first four terms of an arithmetic sequence.

$$4 \quad 12 \quad 20 \quad 28$$

a) Write down the next two terms in the sequence.

.... and ....
[1]

b) The 25th term in the sequence is 196. Write down the 23rd term in the sequence.

....
[1]

c) Thomas says that the 12th term is 90. Explain why he is incorrect.

....

....
[2]

[Total 4 marks]

2 To find the next term in the sequence below, you add together the two previous terms.

Fill in the gaps to complete the sequence.

| 3 | .... | 7 | .... | .... | 29 |

[Total 2 marks]

3 The nth term in a quadratic sequence is given by the formula $n^2 + 3n - 2$.

Write down the first three terms of the sequence.

...................., ... and ....
[Total 2 marks]

4 The nth term of a sequence is given by the rule $3n - 10$.
Two consecutive terms in the sequence have a sum of 223.

Work out the two terms.

The two consecutive terms will be the nth term and the (n + 1)th term. Use the rule for the nth term to find expressions for them.

...................., ...
[Total 4 marks]

Section Two — Algebra

5 This question is about the sequence 3, 7, 11, 15, 19...

a) Find the nth term of the sequence.

....
[2]

b) Explain whether or not 89 is a term in this sequence.

....

....
[2]
[Total 4 marks]

6 The first five terms in a sequence are 10, 40, 90, 160, 250...

a) Find the next two terms in the sequence.

.... , ...
[2]

b) Find an expression for the nth term of the sequence.

....
[2]
[Total 4 marks]

7 The first five terms in a sequence are 2, 10, 30, 68, 130...

a) Find the nth term of the sequence.

Remember — you divide by 6 for cubic sequences, not 3.

....
[2]

b) Using your answer to part a), find an expression for the product of the nth and $(n + 1)$th terms of the sequence. Simplify your answer as much as possible and give your answer without brackets.

....
[3]
[Total 5 marks]

8 The term-to-term rule of a sequence is $u_{n+1} = 2u_n$.

a) If $u_1 = 3$, find the values of the next three terms in the sequence.

.... , ... , ...
[2]

b) Find an expression for the nth term of the sequence.

....
[2]
[Total 4 marks]

9 The patterns below are made up of grey and white squares.

Pattern 1   Pattern 2   Pattern 3   Pattern 4

a) Find an expression for the number of grey squares in the nth pattern.

....
[2]

b) Giles makes two consecutive patterns in the sequence.
He uses 414 grey squares in total. Which 2 patterns has he made?

....
[3]

c) Find an expression for the total number of squares in the nth pattern.

....
[3]
[Total 8 marks]

Score:
37

Algebraic Proportion

1 Sketch the following proportions on the axes below them.

y is proportional to x

y is inversely proportional to x

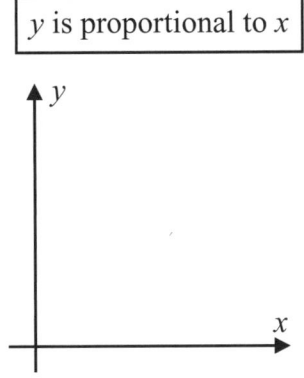
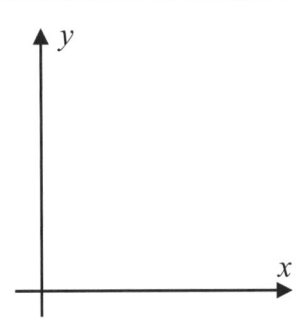

[Total 2 marks]

2 a is inversely proportional to b, where $a = 436$ when $b = 18$.

a) Write an equation where a is expressed in terms of b.

....
[3]

b) Find the value of b when $a = 327$.

$b = $ ....
[1]

[Total 4 marks]

3 The kinetic energy (E joules) of a moving object varies in direct proportion to the square of its velocity (v m/s).

a) A rock falls off a cliff. When its velocity is 24 m/s, its kinetic energy is 1008 joules. Calculate E when $v = 32$.

$E = $ ....
[3]

b) A car's speed increases by 30%. Calculate the percentage increase in its kinetic energy.

.... %
[4]

[Total 7 marks]

Score: 13

Coordinates

1 Two points have been plotted on the grid below. They are labelled **A** and **B**.

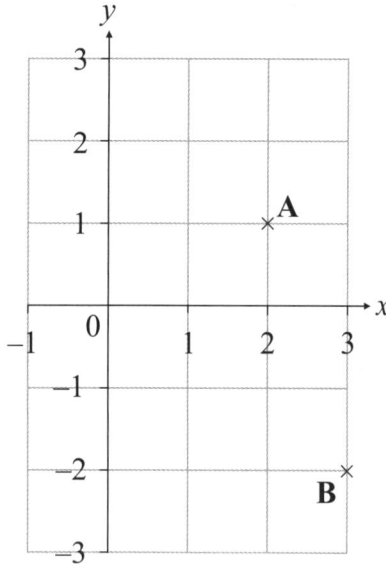

a) Give the coordinates of point **A**.

(............, ...)
[1]

b) Point **C** has the coordinates (1, −1). Mark this point on the grid on the left using a cross (×) and label it **C**.

[1]

c) Points **A**, **B** and **D** form an isosceles triangle when joined together with straight lines. Circle the coordinates that could **not** be the position of point **D**.

(0, −1)   (−1, 0)   (0, 1)   (1, −2)

[1]

[Total 3 marks]

2 Points **A** and **B** have been plotted on the grid below.

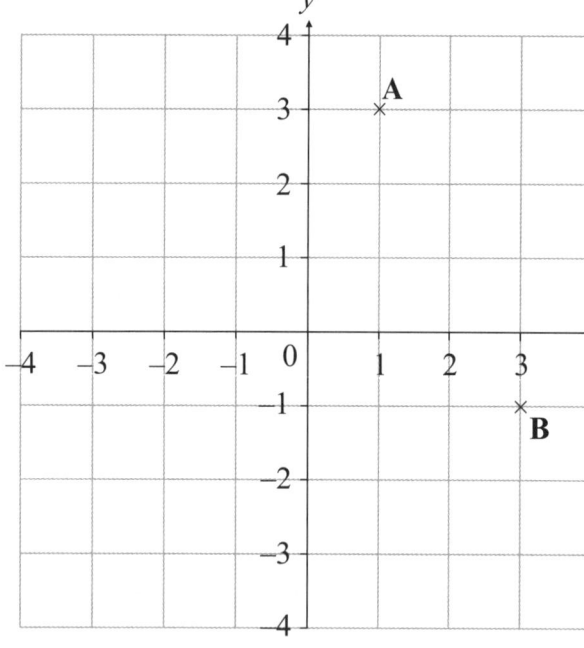

a) Write down the coordinates of the midpoint of the line segment **AB**.

$\left(\dfrac{1+\ldots}{2}, \dfrac{\ldots+\ldots}{\ldots}\right) = (\ldots, \ldots)$

(............, ...)
[2]

b) Point **C** has the coordinates (0, −1). Given that line **AB** and line **CD** have the same midpoint, find the coordinates of point **D**.

(............, ...)
[2]

[Total 4 marks]

Score: ☐ / 7

Straight-Line Graphs

1 Use the grid for the questions below.

a) Draw and label the following lines.

$y = 3$

$x = -2$

$y = x$

[3]

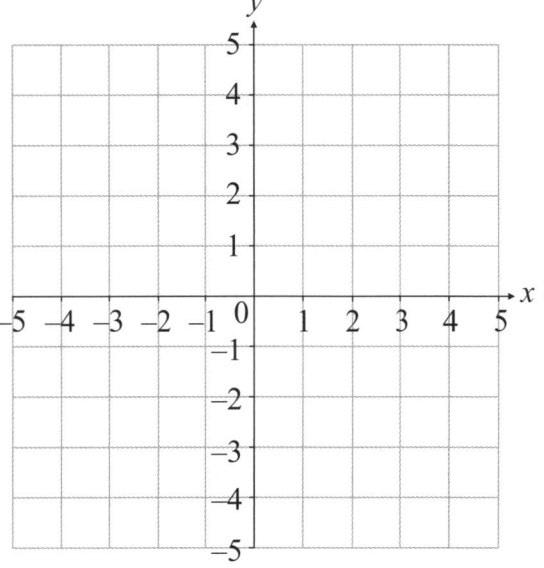

b) What are the coordinates of the point where the lines $y = 3$ and $y = x$ meet?

(..............., ...)

[1]

[Total 4 marks]

2 Answer each question below.

a) Complete this table of values for the equation $y = 3x - 2$.

x	-2	-1	0	1	2
y		-5			

[2]

b) Use your table of values to plot the graph of $y = 3x - 2$ on the grid.

[2]

c) On the same grid, plot the graph of $y = -2x + 2$ from $x = -2$ to $x = 2$.

In the exam, you might not always be given a table of values to help you plot a graph — but it's a good idea to draw your own.

[3]

[Total 7 marks]

Section Three — Graphs, Functions and Calculus

3 The graph below shows 3 lines.

For each line, circle the correct gradient.

a)
    2      $-\frac{1}{2}$ $\frac{1}{2}$ −2

[1]

b)
    −1 $\frac{1}{3}$ 0   1

[1]

c)
    5      $\frac{1}{2}$ −3   −2

[1]
[Total 3 marks]

4 Look at the graph on the right.

Find the equation of the straight line.
Give your answer in the form $y = mx + c$.

Find the gradient: .... = ...

Line crosses y-axis at .... ,

so equation of line is y = ...

....
[Total 3 marks]

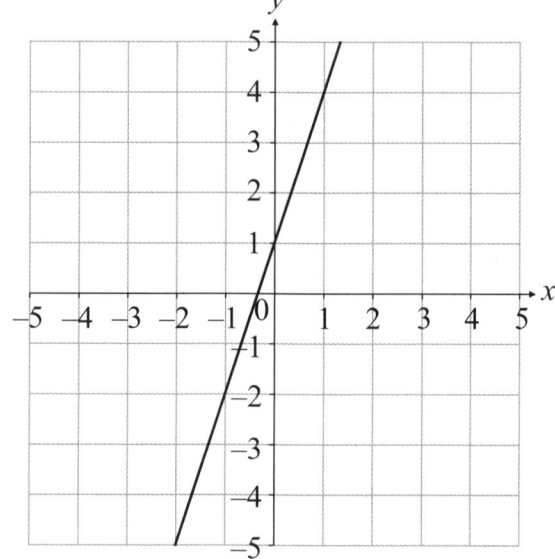

5 Use the grid for the questions below.

a) Draw the line $y = x + 1$.

[2]

b) By drawing a second line, find the equation of the line parallel to $y = x + 1$ which passes through the point (2, 1).
Give your answer in the form $y = mx + c$.

....
[2]
[Total 4 marks]

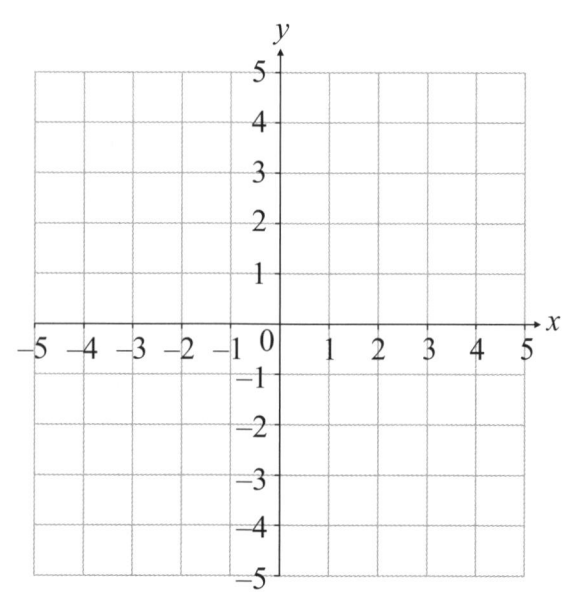

6 $y = 4x - 3$ is the equation of a straight line.

Find the equation of the line parallel to $y = 4x - 3$ that passes through the point $(-1, 0)$.
Give your answer in the form $y = mx + c$.

....
[Total 3 marks]

7 The lines $y = 3x + 4$ and $y = 2x + 6$ intersect at the point M.

Line **N** goes through point M and is perpendicular to the line $y = \frac{1}{2}x + 6$.
Find the equation of line **N**. Give your answer in the form $y = mx + c$.

....
[Total 5 marks]

8 A straight line, **S**, passes through the points (a, b) and (c, d).

It is given that: $2a + 4 = 2c$
$b - 6 = d$

a) What is the gradient of **S**?

Gradient = ...
[3]

b) Line **R** is perpendicular to line **S** and passes through $(6, 3)$. Find the equation of the line.
Give your answer in the form $y = mx + c$.

....
[3]

[Total 6 marks]

Exam Practice Tip
The equation of every straight line can be written in the form y = mx + c, where m is the gradient of the line and c is the y-intercept. If you have to find the equation of a straight-line graph in the exam, you can check your answer by substituting a couple of points on the line back into your equation.

Score: 35

Quadratic Graphs

1 The graph of $y = x^2 + 2x + c$ is shown below.

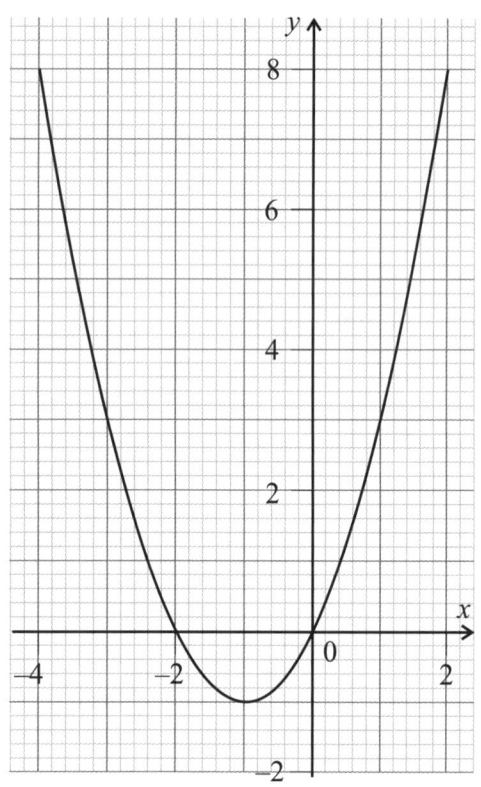

a) Circle the correct value of c.

–2   1   0   –1   –4

[1]

b) Circle both solutions of $x^2 + 2x = 0$.

–1   –2   3   0   1

[1]

c) Write down the coordinates of the lowest point of the curve.

(............, ...)

[1]

d) Circle the correct equation for the line of symmetry of the curve.

$x = -2$ $y = -1$ $x = -1$ $y = 0$

[1]

[Total 4 marks]

2 A table of values for $y = x^2 - 5$ is shown below.

x	–3	–2	–1	0	1	2
y	4	–1	–4	–5	–4	–1

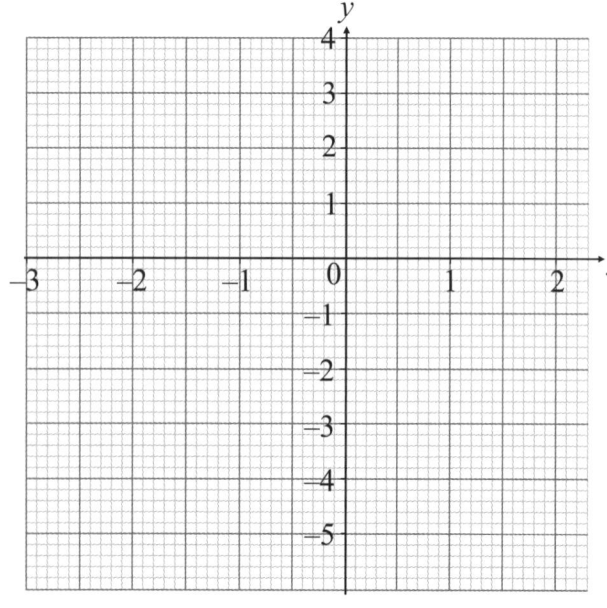

a) Draw the graph of $y = x^2 - 5$ on the grid.

[3]

b) Use your graph to estimate the negative solution of the equation $x^2 - 5 = 0$. Give your answer to 1 decimal place.

x = ...

[1]

[Total 4 marks]

3 This graph below shows $y = x^2 - 3x + a$.

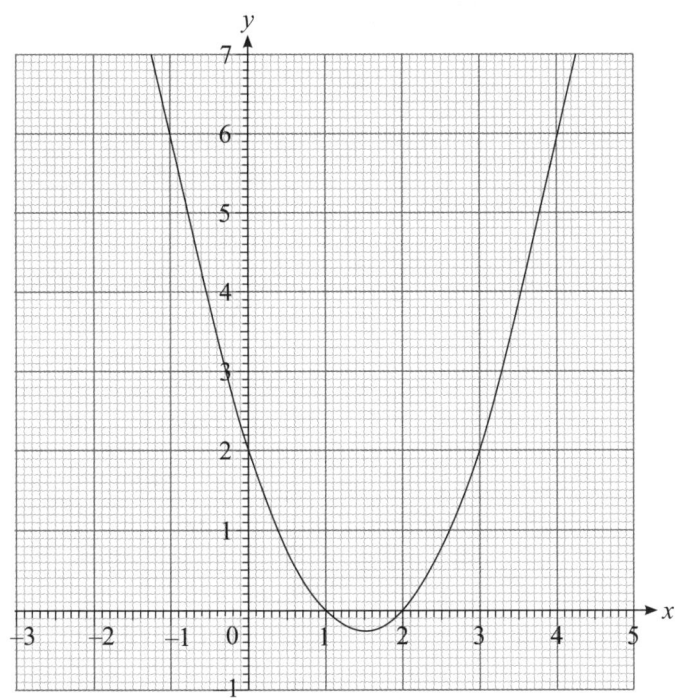

a) Estimate the coordinates of the lowest point on the graph of $y = x^2 - 3x + a$.

(.............. , ...)
[1]

b) Write down the value of a.

a = ...
[1]

[Total 2 marks]

4 The graph of $y = 2x - 2$ is shown on the right.

a) Complete the table of values of $y = -x^2 + 2x + 1$.

x	−2	−1	0	1	2	3	4
y	−7	−2		2	1		−7

[2]

b) Draw the graph of $y = -x^2 + 2x + 1$ for $-2 \leq x \leq 4$ on the grid.

[3]

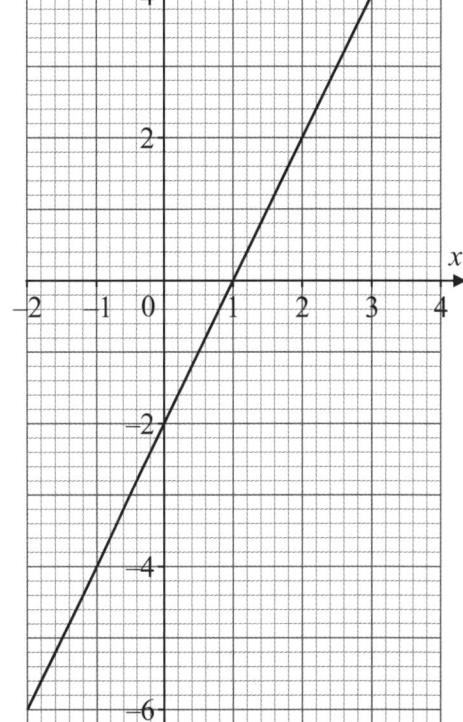

c) Use your graph to estimate the solutions to the following equations. Give your answers correct to 1 decimal place.

i) $-x^2 + 2x + 1 = -3$

x = ... and x = ...
[2]

ii) $-x^2 + 2x + 1 = 2x - 2$

x = ... and x = ...
[2]

[Total 9 marks]

Harder Graphs

1 The graph of $y = \frac{8}{x}$ is a reciprocal graph.

a) Complete the table below for $y = \frac{8}{x}$.

x	1	2	4	8	10	16
y	8	4		1		0.5

[2]

b) On the grid below, draw the graph of $y = \frac{8}{x}$ for $1 \leq x \leq 16$.

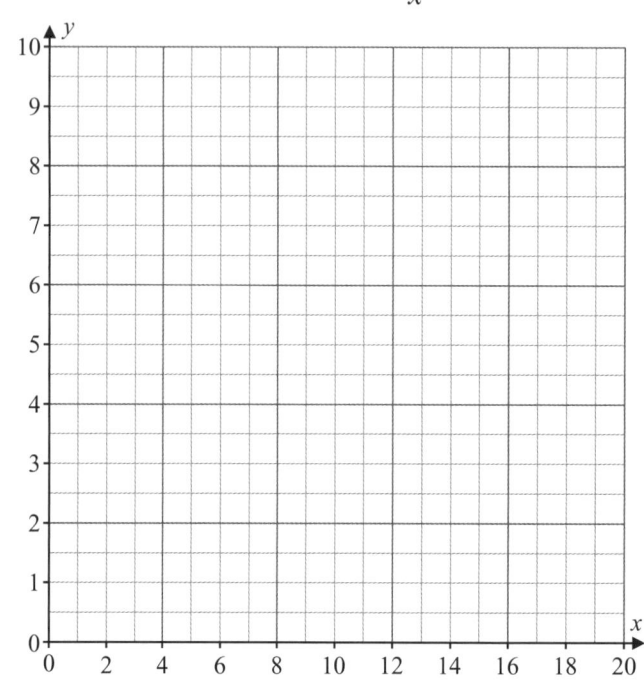

Remember — don't use a ruler to join up the dots in curved graphs.

[3]

c) By drawing the line $y = x + 3$ on the grid, estimate the positive solution to $x + 3 = \frac{8}{x}$. Give your answer to 1 decimal place.

$x = $ ....

[3]

[Total 8 marks]

2 Five functions are given in the box on the right.

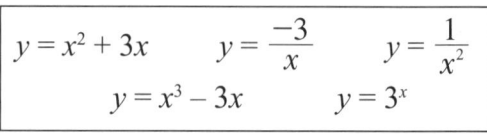

$y = x^2 + 3x \qquad y = \frac{-3}{x} \qquad y = \frac{1}{x^2}$

$y = x^3 - 3x \qquad y = 3^x$

Match each graph with its correct function from the box.

a)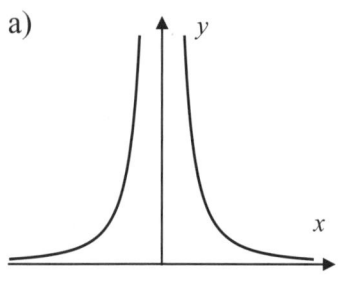

b)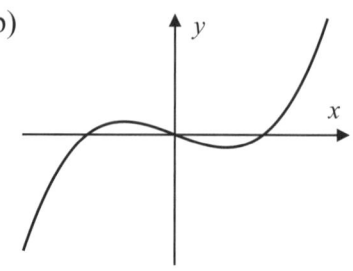

c)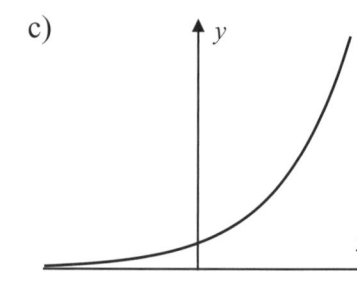

$y = $ .... $\qquad$ $y = $ .... $\qquad$ $y = $ ....

[Total 3 marks]

3 This question is about the function $y = x^3 - 4x^2 + 4$.

a) Complete the table below.

x	−1	−0.5	0	0.5	1	1.5	2	2.5	3	3.5	4
y	−1	2.875	4	3.125	1	−1.625	−4	−5.375		−2.125	

[2]

b) Use your table to draw the graph of $y = x^3 - 4x^2 + 4$ on the grid, for values of x in the range $-1 \leq x \leq 4$.

[4]

c) Estimate the solutions of the equation $x^3 - 4x^2 + 4 = 0$ for $-1 \leq x \leq 4$.
Give your answers to 1 decimal place.

x = ...

x = ...

x = ...

[3]

[Total 9 marks]

4 Over the next five years, the estimated value, V, of a bracelet can be modelled by the equation $V = 2^n + 10$, where n is the number of years.

a) Complete the table and draw the graph of $V = 2^n + 10$ for $0 \leq V \leq 5$ on the grid below.

x	y
0	
1	12
2	14
3	18
4	
5	42

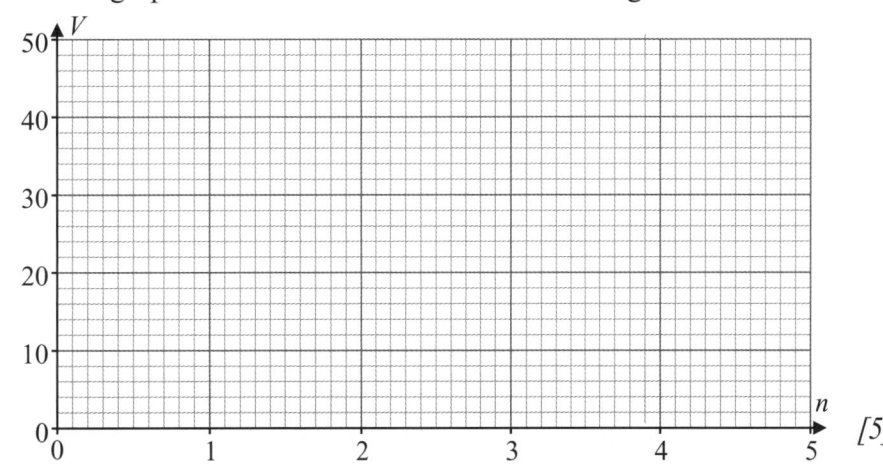

[5]

b) Estimate how long it will be before the bracelet is worth over $30.
Give your answer to 1 decimal place.

.... years

[1]

[Total 6 marks]

Real-Life Graphs

1 The table shows how much petrol a sports car used on 3 journeys.

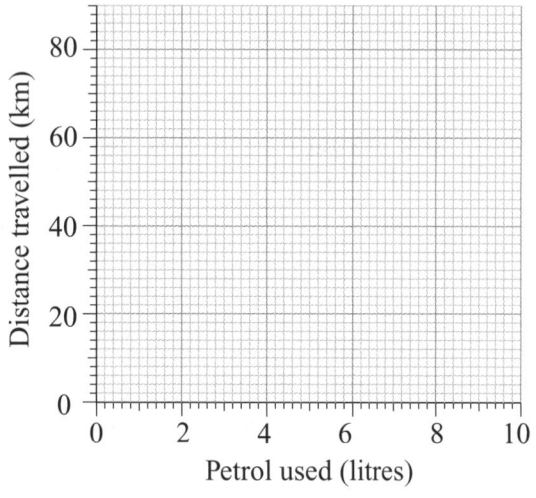

Distance (km)	16	40	72
Petrol used (litres)	2	5	9

a) Using the table above, draw a conversion graph on the grid to the left.
[2]

b) Use your graph to estimate how much petrol the sports car would use on a 28 km journey.

.... litres
[1]

c) Find the gradient of the line.

....
[2]

d) What does the gradient of the line represent?

....
[1]
[Total 6 marks]

2 Each of the vessels below is filled with water at a constant rate.

  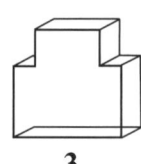 

   1   2   3   4

Each of these graphs shows the depth of water within a vessel in relation to time.

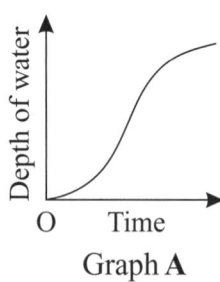 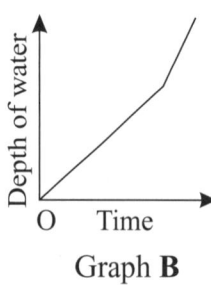 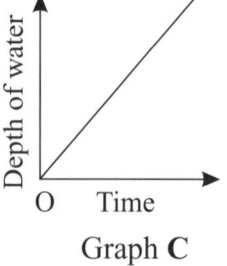 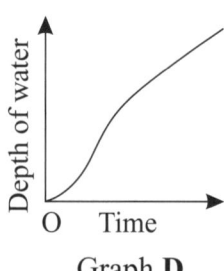

  Graph **A**   Graph **B**   Graph **C**   Graph **D**

Match the vessel with the correct graph.

  Graph **A** and .... Graph **B** and .... Graph **C** and .... Graph **D** and ....
[Total 2 marks]

Score:

8

Section Three — Graphs, Functions and Calculus

Travel Graphs

1 The distance/time graph on the right shows Nish's journey from his house (**A**) to the zoo (**C**), which is 10 km away.

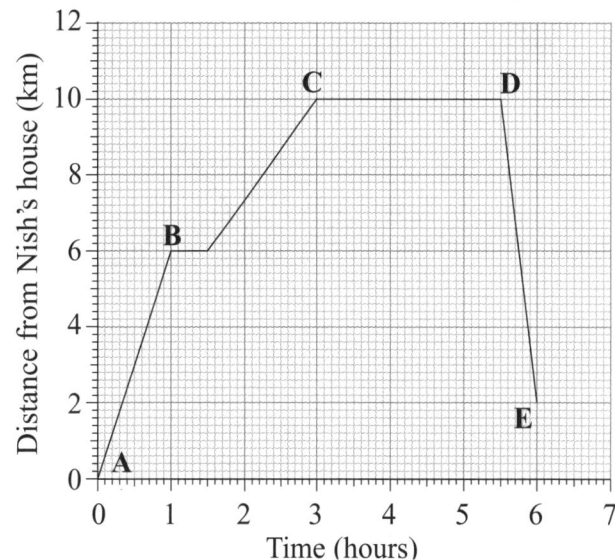

a) Calculate Nish's speed between **A** and **B**.

.... km/h
[1]

b) How long did Nish spend at the zoo?

.... hours
[1]

c) After the zoo, Nish stopped at the shops (**E**) for 30 minutes before walking straight home at a constant speed. Given that he arrived home 7 hours after he left, complete the graph above.
[2]

[Total 4 marks]

2 The speed of a motorcycle is recorded over a minute.

a) Draw a speed-time graph using the following information:
The motorcycle sets off from a standstill, accelerating at a constant rate for 10 seconds until it is moving at 10 m/s. It moves at a constant speed for the next 20 seconds. The motorcycle then accelerates at a constant rate for 7 seconds until it is moving at 24 m/s. It moves at the same speed for 15 seconds before decelerating until it stops after 8 seconds.

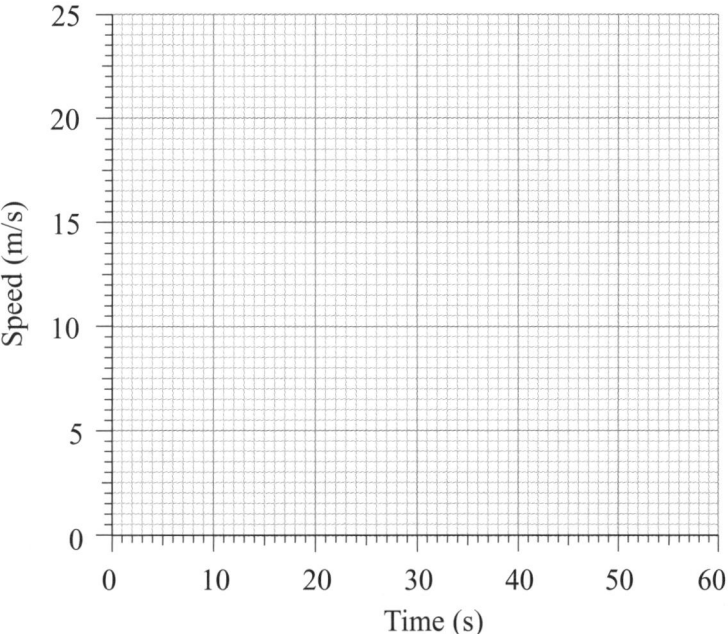

[3]

b) Calculate the acceleration of the motorcycle at 35 seconds.

The gradient of the graph tells you the acceleration.

.... m/s²
[1]

[Total 4 marks]

3 The speed-time graph below shows the speed of a ferry on a 5-hour journey.

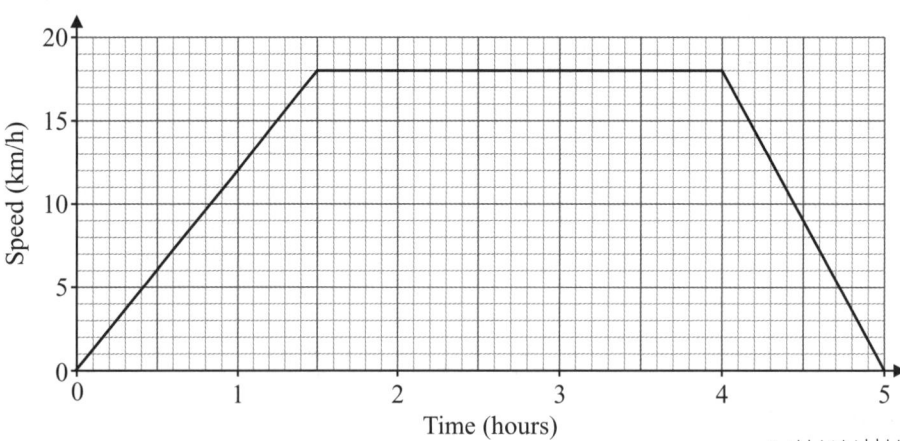

a) Calculate the total distance travelled by the ferry on this journey.

The total distance is given by the area under the graph.

.... km
[2]

b) Calculate the ferry's average speed on this journey.

.... km/h
[1]
[Total 3 marks]

4 The graph below shows how the speed of a moving vehicle changes over time.

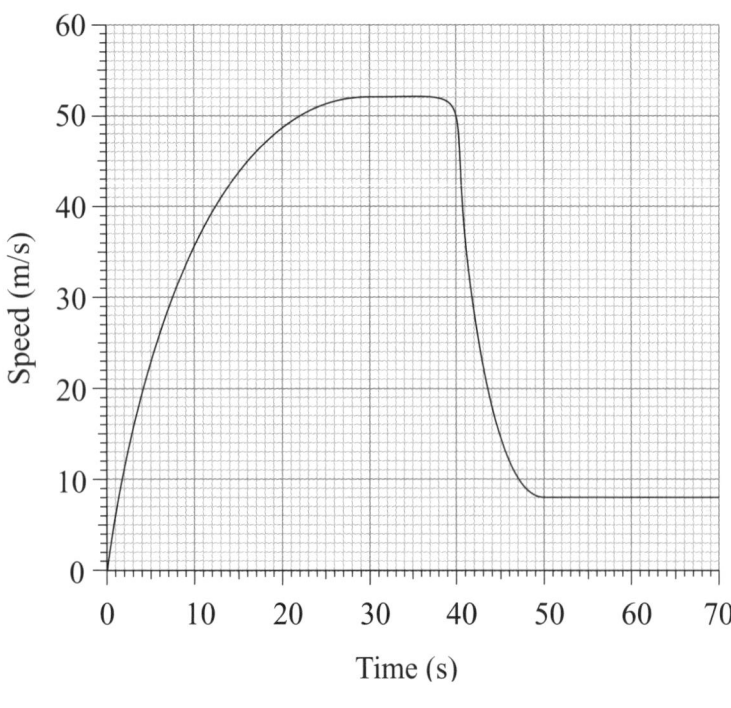

a) Find the average acceleration of the vehicle between 5 seconds and 25 seconds.

.... m/s²
[2]

b) Estimate the acceleration of the vehicle at 45 seconds.

.... m/s²
[3]
[Total 5 marks]

Score:

16

Section Three — Graphs, Functions and Calculus

Functions

1 h is a function defined by h: $x \to \dfrac{6x-5}{2}$

a) Find h(5)

.... [1]

b) Find $h^{-1}(x)$

$h^{-1}(x) =$ .... [2]

c) The domain of h(x) is {2, 3, 4, 5}.

Find the range of h(x). Give the elements of the set as decimals.

{.....................................}
[2]
[Total 5 marks]

2 f and g are functions such that $f(x) = 2x^2 + 3$ and $g(x) = \sqrt{2x-6}$, where $x \geq 3$.

a) Find g(21)

$g(21) = \sqrt{\text{.....................}} = \sqrt{\text{...............}} = \text{...............}$ .... [1]

b) Find gf(x)

Remember to do the function closest to x first.

$gf(x) =$ .... [2]

c) Solve fg(a) = 7

a = ... [3]
[Total 6 marks]

Exam Practice Tip

Make sure you're familiar with the two different ways to write functions, as either one could come up on your exam — for example, f(x) = x + 2 and f: x → x + 2 mean the same thing. It's also worth remembering that the order that functions are written matters a lot — in general, fg(x) is not equal to gf(x).

Score

11

Section Three — Graphs, Functions and Calculus

Differentiation

1 A curve has equation $y = 2x^3 + 4x^2 - 9$.

Find $\frac{dy}{dx}$.

$\frac{dy}{dx} = $ ....
[Total 2 marks]

2 A quadratic curve has the equation $y = -x^2 + 3$.

a) Find $\frac{dy}{dx}$.

$\frac{dy}{dx} = $ ....
[1]

b) Find the gradient of the graph of $y = -x^2 + 3$ at $x = -1$ and $x = 2$.

Gradient at $x = -1$: ....

Gradient at $x = 2$: ....
[2]
[Total 3 marks]

3 The curve with equation $y = 3x^2 - 3x + 4$ has one turning point.

a) Find $\frac{dy}{dx}$.

$\frac{dy}{dx} = $ ....
[2]

b) Find the coordinates of the turning point of the curve with the equation $y = 3x^2 - 3x + 4$.

(................ , ...)
[3]
[Total 5 marks]

4 A curve has equation $y = -x^3 - \frac{21}{2}x^2 + 24x$.

a) Find the coordinates of the two turning points of the curve.

(................ , ...) and (................ , ...)
[6]

b) Determine whether each of the turning points is a maximum or a minimum.
Give reasons for your answers.

[3]
[Total 9 marks]

5 The velocity, v m/s, of an object after t seconds is given by $v = -2t^3 + 9t^2 + 39$ for $0 \leq t \leq 5$.

The maximum velocity occurs at a stationary point on the graph of $v = -2t^3 + 9t^2 + 39$.
Find the maximum velocity of the object over this time.

.... m/s
[Total 6 marks]

Geometry

1 AB is a straight line.

Work out the size of the angle marked *x*.

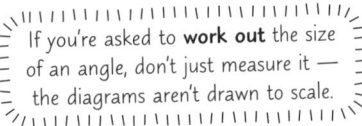

*If you're asked to **work out** the size of an angle, don't just measure it — the diagrams aren't drawn to scale.*

Not to scale

.............................°

[Total 1 mark]

2 Look at the diagram on the right.

Explain why at least one angle must have been labelled incorrectly.

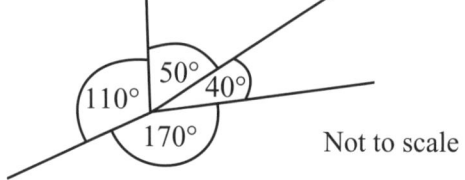

Not to scale

....

[Total 2 marks]

3 BCDE is a quadrilateral. Angle CDE is a right angle.
AF is a straight line.

Work out the size of the angle marked *x*.

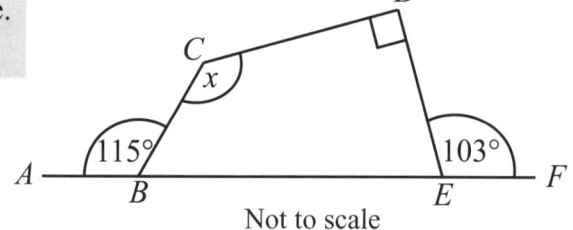

Not to scale

.............................°

[Total 3 marks]

4 ABC is an isosceles triangle.
AB = BC.
AD is a straight line.

Work out the size of angle BCD.

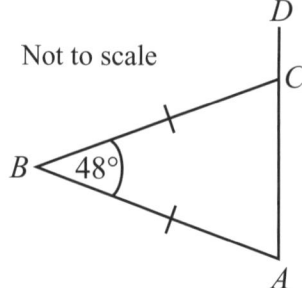

Not to scale

The dashes on the diagram mean that AB is the same length as BC.

.............................°

[Total 3 marks]

5 AB and CD are parallel lines. EF and GH are straight lines.

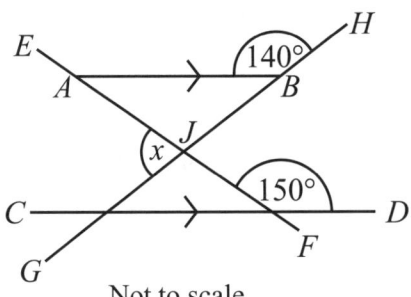

Work out the size of angle x.
Give reasons for each stage of your working.

.............................°
[Total 4 marks]

6 WXYZ is a quadrilateral.
Angle WXY is a right angle.
WXZ and XYZ are triangles.

Work out the size of angle ZXY.

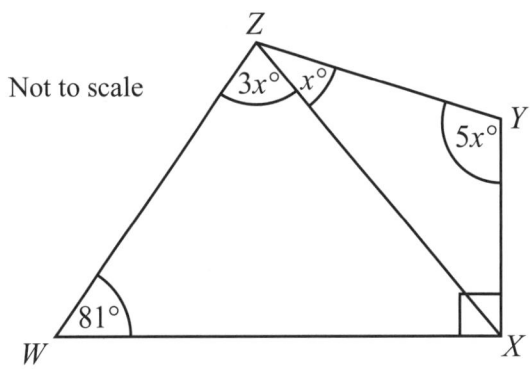

.............................°
[Total 4 marks]

7 BD and JH are parallel straight lines.
AI and EK are straight lines.

 Show that triangle EFG is isosceles.

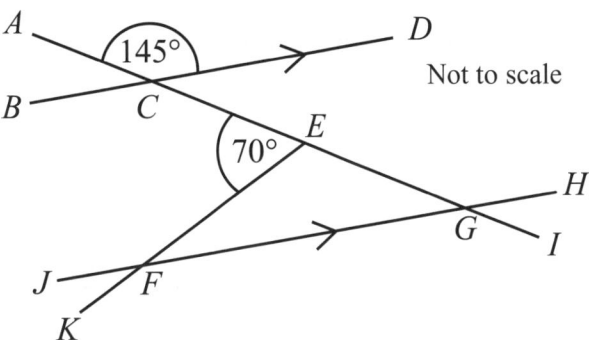

[Total 3 marks]

Exam Practice Tip
If you can't see how to find the angle you've been asked for, try finding other angles in the diagram first — chances are you'll be able to use them to find the one you need. You'll probably have to use a few of the angle rules to get to the answer — if you get stuck, just try each rule until you get to one that you can use.

Score

/20

Section Four — Geometry and Measures

Bearings and Scale Drawings

1 Douglas drew a scale drawing of one of the rooms in his house.

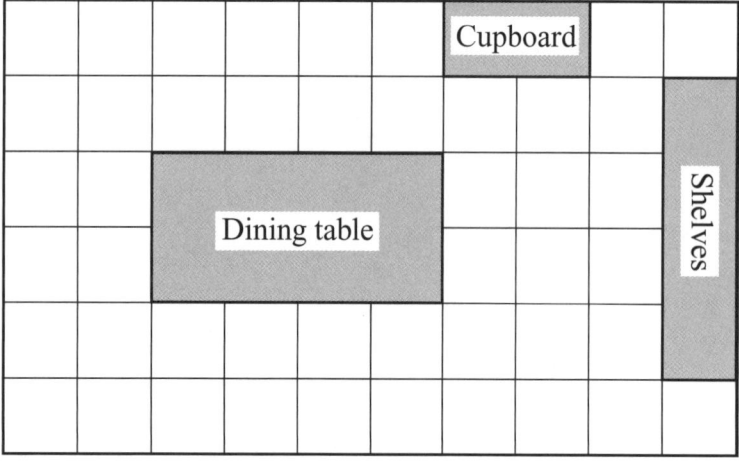

a) His dining table is 2 m long. What is the scale of this drawing?

1 cm to .... m
[1]

b) Work out the real distance from the dining table to the shelves.

.... m
[1]

c) Douglas wants to put a chair measuring 1 m × 1.5 m in the room so that there is a space of at least 0.5 m around it. Is this possible with the current layout of the room? Give a reason for your answer.

....
[2]
[Total 4 marks]

2 The instructions on a treasure map say "start at the cross and walk 400 metres on a bearing of 150°. Then walk 500 metres on a bearing of 090° to find the treasure."

Using a scale of 1 cm = 100 m, accurately draw the path that must be taken to find the treasure on the map below.

Make sure you draw the north line accurately for the second bearing.

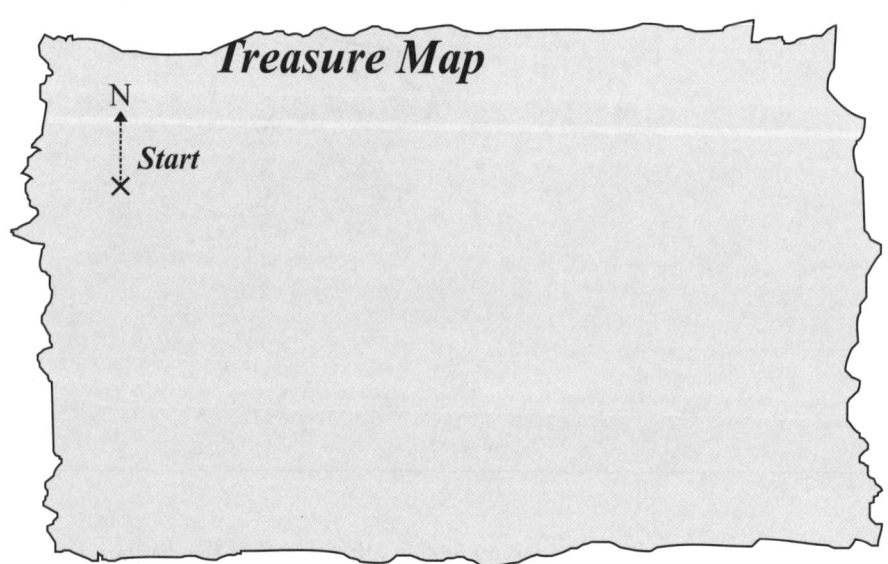

[Total 3 marks]

Section Four — Geometry and Measures

3 Two ships leave a port at the same time.
Ship *A* travels due west for 40 km. Ship *B* travels 60 km on a bearing of 110°.

a) Using a scale of 1 cm = 10 km, draw the journeys of the two ships in the space below and clearly mark their final positions.

N

Port

[3]

b) Measure the final bearing of Ship *B* from Ship *A*.

.............°

[1]

c) Calculate the final bearing of Ship *A* from Ship *B*.

.............°

[2]

[Total 6 marks]

4 The diagram shows the position of two villages, *A* and *B*.

a) A walker hikes from village *A* on a bearing of 035°.
After an hour's walk he stops when village *B* is directly east of his position.
Mark the walker's position on the diagram with a cross (×) and label it *W*.

[2]

b) Another village, *C*, is on a bearing of 115° from village *A*, and on a bearing of 235° from village *B*. Mark the location of village *C* with a cross (×) and label it *C*.

[3]

c) Use a protractor to measure the bearing that the walker must hike on from his position at *W*, in order to reach village *C*.

.............°

[1]

[Total 6 marks]

Score:

19

Section Four — Geometry and Measures

Polygons and Symmetry

1 The diagram below shows a parallelogram.

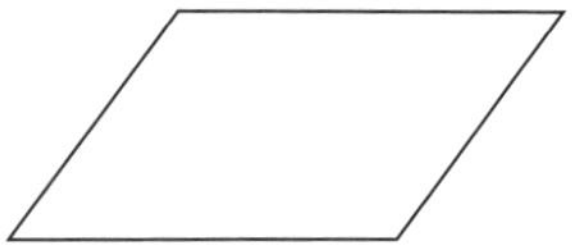

a) How many lines of symmetry does a parallelogram have?

....
[1]

b) What order of rotational symmetry does a parallelogram have?

....
[1]
[Total 2 marks]

2 Find the size of the exterior angle of a regular pentagon.

.............................°
[Total 2 marks]

3 Here are the names of five shapes.

| Trapezium | Isosceles Triangle | Kite | Scalene Triangle | Rhombus |

a) Which shape has the most lines of symmetry?

....
[1]

b) How many of the shapes always have at least one pair of equal angles?

....
[1]
[Total 2 marks]

4 A regular hexagon is split into 4 triangles, as shown in the diagram.

Explain how this shows that the sum of the interior angles in a regular hexagon is 720°. Do not measure any angles.

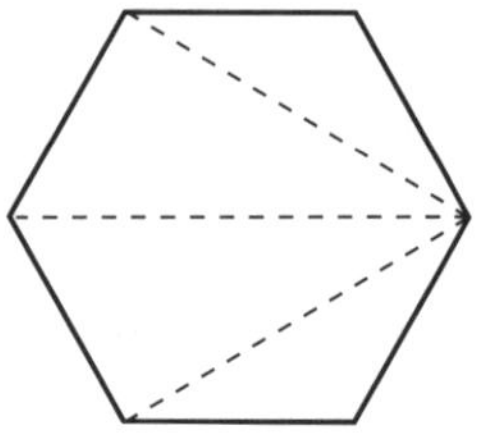

....

....
[Total 1 mark]

Section Four — Geometry and Measures

5 The diagram shows a regular octagon. *AB* is a side of the octagon and *O* is its centre.

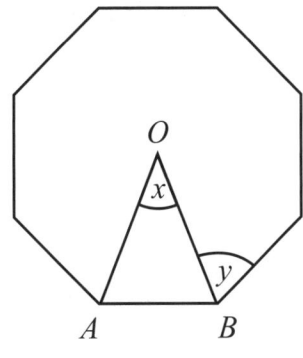

Not to scale

a) Work out the size of the angle marked *x*.

x = ...°
[2]

b) Work out the size of the angle marked *y*.

y = ...°
[2]

[Total 4 marks]

6 The diagram below shows a regular hexagon inside a regular octagon. Vertices *A* and *B* coincide with vertices *I* and *J* respectively.

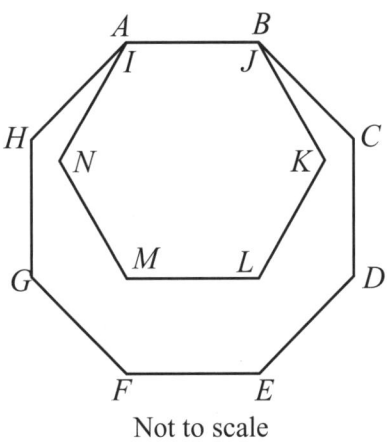

Not to scale

Find the size of angle *CBK*.

.............................°
[Total 3 marks]

7 The polygon on the right has one line of symmetry, shown by the dashed line.

Find the size of the reflex angle marked *x*.

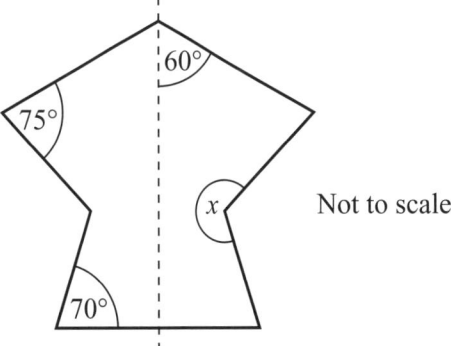

Not to scale

.............................°
[Total 4 marks]

Score:

18

Circle Geometry

1 The line AC is a tangent to the circle, centre O, at B. Angle AOB is 62°.

   Find the size of angle OAB.

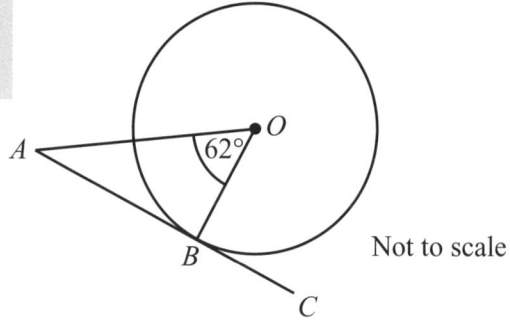

Not to scale

.... °

[Total 2 marks]

2 In the diagram below, points A, B and C lie on the circumference of the circle. The line AB passes through the centre of the circle, O, and the chords AC and BC have the same length.

   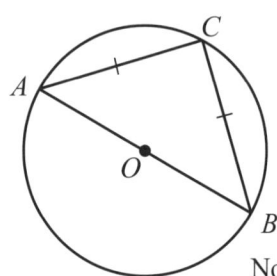   Find the size of angle CBA.

Not to scale

.... °

[Total 2 marks]

3 The diagram on the right shows two circles. The smaller circle passes through the centre, O, of the larger circle, and the circles cross at point X. The line AC passes through the centres of both circles, and crosses the larger circle at B.

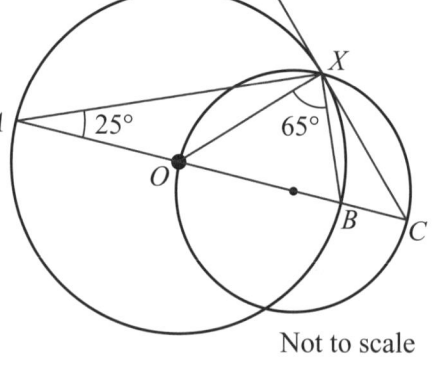

The line CD passes through the point X.

a) Explain how you know that the line CD is a tangent to the larger circle at X.

Start by finding angle OXC.

....

....

....

[2]

Not to scale

Angle XAO = 25° and angle OXB = 65°.

b) Work out the size of angle ACX.

Angle AXO = ...

Angle AXC = ...

Angle ACX = ...

.... °

[3]

[Total 5 marks]

Section Four — Geometry and Measures

4 The diagram shows a circle with centre O. A, B and C are points on the circumference. AD and CD are tangents to the circle and ABE is a straight line. Angle CDO is 24°.

a) Find the size of angle DOC.

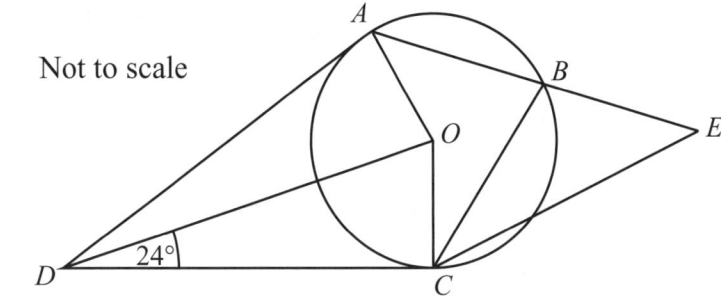

..................°
[2]

b) Find the size of angle CBE.

..................°
[3]

[Total 5 marks]

5 In the diagram below, A, B, C and D are points on the circumference of the circle. EDB is a straight line and FG is the tangent to the circle at point B.
Angle FBD is 102° and angle EDC is 147°.

Find the size of angle CAD.

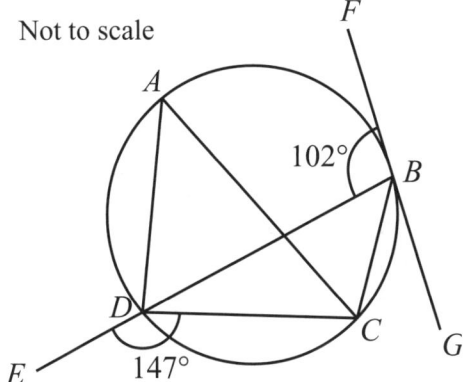

..................°
[Total 4 marks]

6 A, B, C and D are points on the circumference of the circle shown below.

 Show that X is not the centre of the circle.

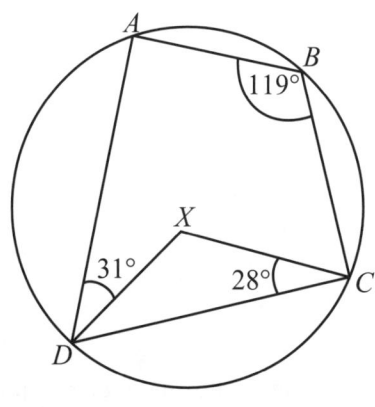

Not to scale

[Total 3 marks]

7 A, B, C and D are points on the circumference of the circle with centre O.
FE is the tangent to the circle at D and angle BDE = 53°.

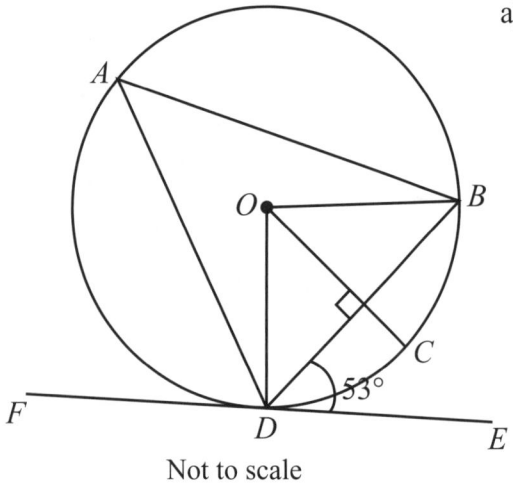

Not to scale

a) Find the size of angle DOB.

.............................°
[2]

b) Explain why angle COB is half the size of angle DOB.

....

....

....
[2]
[Total 4 marks]

8 In the diagram, A, B, C and D are points on the circle, centre O. The chords AB and CD have the same length. The points E and F are on the chords, where OE is perpendicular to AB, and OF is perpendicular to CD. The points OEF form a triangle as shown.

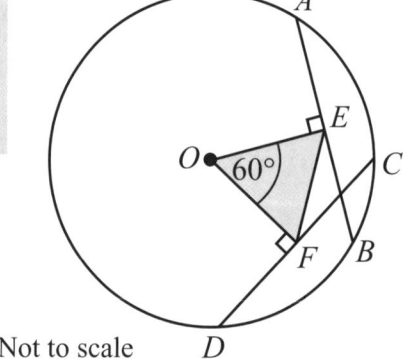

Not to scale

Given that angle EOF is 60°, show that this triangle is equilateral.

[Total 4 marks]

Exam Practice Tip
Make sure you know the rules about circles really, really well. Draw them out and stick them all over your bedroom walls, your fridge, even your dog. Then in the exam, go through the rules one by one and use them to fill in as many angles in the diagram as you can. Keep an eye out for sneaky isosceles triangles too.

Score: **29**

Congruence and Similarity

1 The diagram on the right shows eight quadrilaterals, labelled **A** to **H**.

   Which two pairs of quadrilaterals are congruent?

   .... and ....

   .... and ....
   [Total 2 marks]

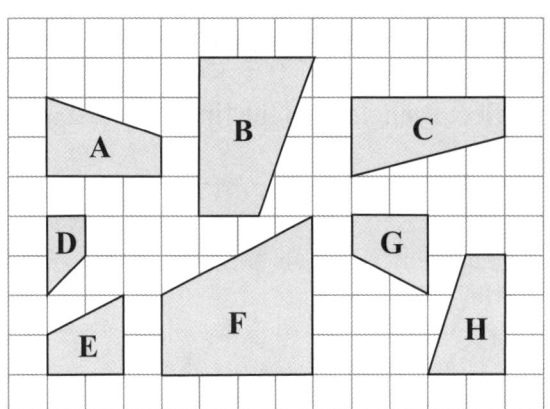

2 Triangles *ABC* and *DBE* are similar. *ABE* and *CBD* are both straight lines.

   Find the missing values *x* and *y*.

   Hint: You need to use the rule about vertically opposite angles to answer this question.

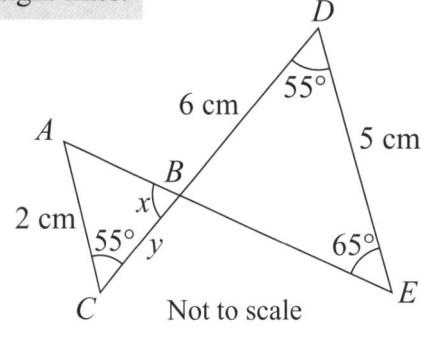

   $x = $ .... °

   $y = $ .... cm
   [Total 3 marks]

3 The diagram shows the design for a tabletop, made by cutting off the end of a triangular piece of wood to form a trapezium.

   Calculate the value of *x*.

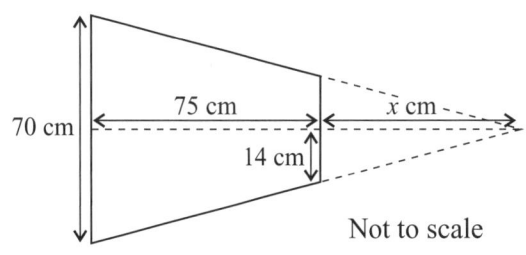

   .... cm
   [Total 3 marks]

Exam Practice Tip
Similar shapes are basically just enlargements of each other (though they could be rotated or flipped), so the first step is usually to find the scale factor of the enlargement. Make sure you know which sides correspond to which — especially if one of the shapes has been flipped over, which can make it harder to keep track.

Score / 8

The Four Transformations

1 Triangle **A** has been drawn on the grid below.

Reflect triangle **A** in the line $x = -1$. Label your image **B**.

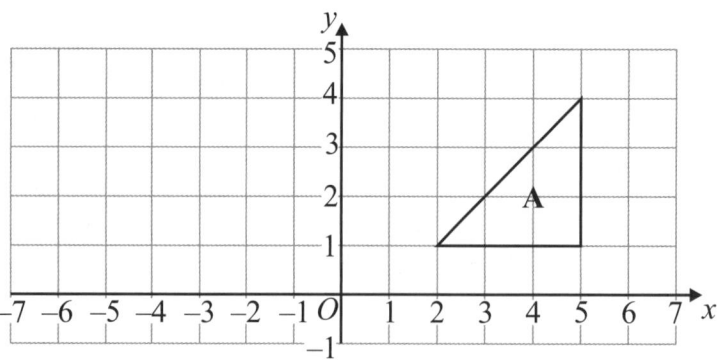

[Total 2 marks]

2 On the grid below, enlarge the triangle by a scale factor of 3, centre (–4, 0).

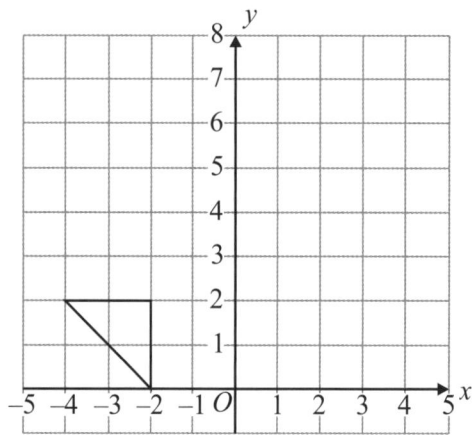

[Total 2 marks]

3 The diagram below shows shape **A**.

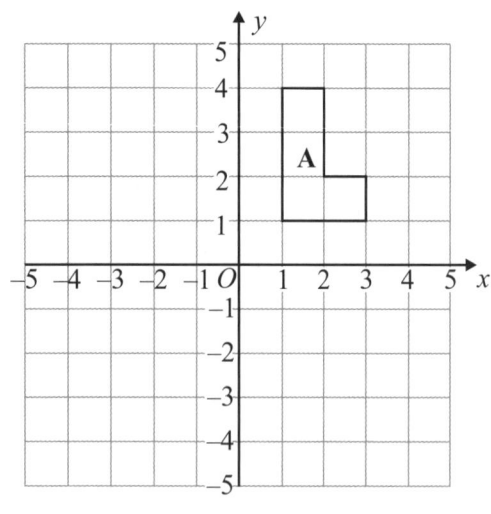

a) Rotate shape **A** by 180° about (0, 0). Label the image **B**.

[2]

b) Translate shape **B** by the vector $\begin{pmatrix} 4 \\ 2 \end{pmatrix}$. Label the image **C**.

[2]

c) Describe fully the single transformation which maps shape **A** onto shape **C**.

....

....

[3]

[Total 7 marks]

4 Triangles **A** and **B** have been drawn on the grid on the right.

a) Describe fully the single transformation which maps shape **A** onto shape **B**.

....

....

....
[2]

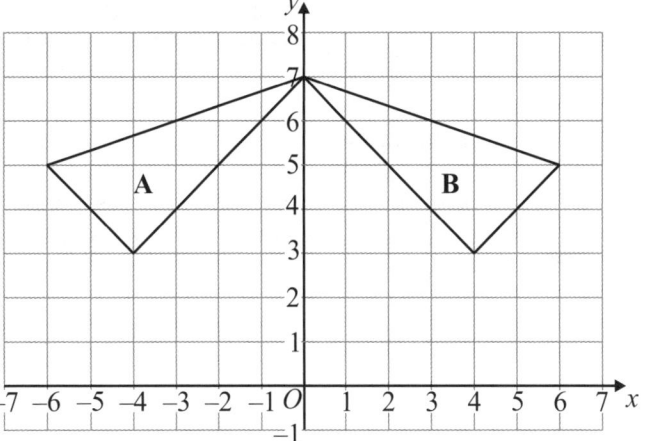

b) Enlarge triangle **B** by scale factor $\frac{1}{2}$ with centre of enlargement (–6, 1). Label your image **C**.

[2]

[Total 4 marks]

5 Triangles **R** and **S** have been drawn on the grid below.

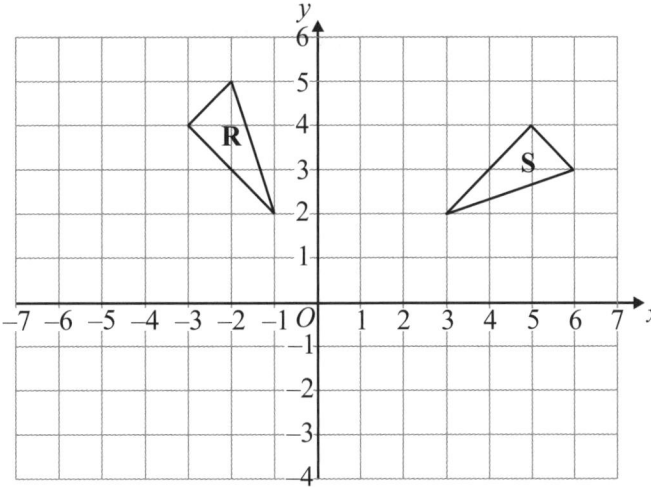

a) Describe fully the single transformation which maps shape **R** onto shape **S**.

....

....
[3]

b) Reflect triangle **R** in the line $y = x$ and then enlarge it with centre (1, –2) and scale factor –2. Label the resulting shape **T**.

[4]

[Total 7 marks]

Exam Practice Tip

Make sure you give all the details when you describe a transformation. If a question is worth three marks, you'll probably need to give the type of transformation, plus two other bits of information — e.g. the scale factor and the centre of enlargement for enlargements, and the centre and the angle (including direction) for rotations.

Score

22

Section Four — Geometry and Measures

More Enlargements

1 The radius of a tennis ball and the radius of a basketball are in the ratio 1 : 7.

Assuming both balls are spheres, work out the ratio of the volume of a tennis ball to the volume of a basketball.

For an enlargement with scale factor n, the scale factor of the volumes will be n^3.

....
[Total 1 mark]

2 A parallelogram has an area of 7 cm².

The parallelogram is enlarged with scale factor 3. Work out the area of the enlarged parallelogram.

.... cm²
[Total 2 marks]

3 Cuboids A and B are similar. Cuboid A has a height of 1.5 cm and the area of its top face is 12.5 cm². The scale factor of enlargement from A to B is $\frac{8}{5}$.

Find the volume of cuboid B.

.... cm³
[Total 2 marks]

4 Cylinder B is an enlargement of cylinder A.
The ratio of the volume of cylinder A to the volume of cylinder B is 27 : 64.
The surface area of cylinder A is 81π cm².

Find the surface area of cylinder B.

.... cm²
[Total 3 marks]

Score: ☐
8

Unit Conversions

1 The playing surface of a snooker table has an area of 39 200 cm².

Convert the area of the snooker table into m².

.... m²
[Total 2 marks]

2 A barrel has a capacity of 150 litres.

What is the capacity of the barrel in cubic metres?

150 litres = ... = ... cm³

1 m³ = ... = ... cm³

So .... cm³ = ...

= .... m³   .... m³
[Total 3 marks]

3 Aled is going on a road trip from the US to Canada. He can buy gasoline from either country on his trip. A gas station in the US sells gasoline for 0.80 US dollars per litre, while one in Canada sells it for 1.15 Canadian dollars per litre.

How much cheaper is it (in US dollars) to buy 40 litres of gasoline in the US rather than Canada? Use the conversion of 1 Canadian dollar = 0.75 US dollars.

.... US dollars
[Total 3 marks]

4 A large cube has side lengths of 3 m. It is filled up with small cubes each with a side length of 60 mm.

How many of the smaller cubes will fit inside the large cube?

....
[Total 3 marks]

Score:
11

Section Four — Geometry and Measures

Time

1 Part of the bus timetable from Coventry to Rugby is shown on the right.

Coventry	14 45	16 15	17 45
Bubbenhall	–	16 40	18 10
Birdingbury	–	17 04	–
Rugby	15 35	17 30	18 40

a) Lisa arrives at Birdingbury bus stop at 16 58. How long will she have to wait for the bus to Rugby?

The dashes on the timetable mean the bus doesn't stop.

.... minutes
[1]

The 16 15 bus from Coventry continues to Lutterworth after Rugby. It arrives in Lutterworth at 18 15.

b) If Anne catches this bus from Bubbenhall, how long will it take her to get to Lutterworth?

.... hours .... minutes
[1]
[Total 2 marks]

2 Naveed took a flight from the UK to India. It was 15 50 in the UK when his flight took off and 05 45 in India when his flight landed.

Given that India is $5\frac{1}{2}$ hours ahead of the UK, work out how long the flight lasted. Give your answer in hours and minutes.

.... hours .... minutes
[Total 2 marks]

3 A cake has to be baked for 2¼ hours plus 10 minutes for every 100 g the cake weighs.

a) Mary put a 400 g cake in the oven at 9:55 am. What time should she take the cake out of the oven?

....
[3]

b) Mac baked his cake from 10:45 to 14:02. How much did his cake weigh?

.... g
[3]
[Total 6 marks]

Score: 10

Section Four — Geometry and Measures

Speed, Density and Pressure

1 John and Alan hired a van. Their receipt gave them information about how much time they spent travelling in the van, and how fast they went.

> Travelling time: 1 hour 15 minutes
> Average speed: 56 km/h

Calculate the distance that John and Alan travelled in the van.

.... km
[Total 2 marks]

2 Hannah hits a wall with a sledgehammer, exerting a force of 1000 newtons. The head of the sledgehammer that makes contact with the wall has an area of 50 cm².

Calculate the pressure exerted by the sledgehammer on the wall. [Pressure = $\frac{\text{Force}}{\text{Area}}$]

.... N/cm²
[Total 1 mark]

3 A metal alloy is made up of 120 g of metal A and 130 g of metal B. Metal A has a density of 6 g/cm³ and metal B has a density of 5 g/cm³.

a) What is the total volume of metal used in the alloy? [Density = $\frac{\text{Mass}}{\text{Volume}}$]

Hint: find the volume of each metal separately.

.... cm³
[3]

b) What is the density of the alloy? Give your answer to 1 decimal place.

.... g/cm³
[2]
[Total 5 marks]

4 The population density of a city is 8770 people/km².

How many people would you expect to live in a 24 000 m² area of this city? Give your answer to the nearest whole number. [Population density = $\frac{\text{People}}{\text{Area}}$]

Hint: start by converting the area to km².

.... people
[Total 3 marks]

Section Four — Geometry and Measures

5 Brass is a metallic alloy. One type of brass consists only of copper and zinc in the ratio 7:3 by volume. Copper has a density of 8.9 g/cm³ and zinc has a density of 7.1 g/cm³.

What is the density of this type of brass? [Density = $\frac{\text{Mass}}{\text{Volume}}$]

10 cm³ of brass contains .... cm³ of copper and .... cm³ of zinc.

.... cm³ of copper has a mass of .... g

.... cm³ of zinc has a mass of .... g

10 cm³ of brass has a mass of .... g

Density of brass = ... g/cm³

.... g/cm³
[Total 4 marks]

6 To get to school, Sumire walks to the bus stop and then takes the bus. The bus stop is 1.2 km from her house, and the bus journey to her school is 32 km. The bus travels with an average speed of 40 km/h.

One day, Sumire's journey took 1 hour and 8 minutes. Given that she waited 5 minutes for the bus, calculate her average walking speed to the bus stop.

.... km/h
[Total 3 marks]

7 The cone below has a base diameter of $20x$ cm. When the base of the cone rests on horizontal ground, it exerts a pressure of 650 N/m² because of its weight force.

a) Calculate the weight of the cone in terms of x and π. [Pressure = $\frac{\text{Force}}{\text{Area}}$]

.... N
[3]

b) The diameter of the cone is halved but the weight is kept the same. What effect will this have on the pressure exerted on the ground?

[2]
[Total 5 marks]

Exam Practice Tip
Converting between different compound units, such as units of speed, can be tricky. Just remember that they're made up of two measures (e.g. distance and time for speed). If the units of both measures are changing (e.g. km/h to m/s), then you'll need to do two conversions — one for each (e.g. km to m, and h to s).

Score: /23

Perimeter and Area

1 The diagram on the right shows an isosceles trapezium.

Find the area of the trapezium.

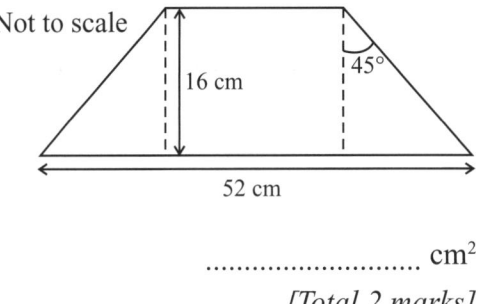

.... cm²
[Total 2 marks]

2 A letter "O" is formed by cutting a circle of diameter 6 cm from the centre of a circular piece of card with diameter 10 cm.

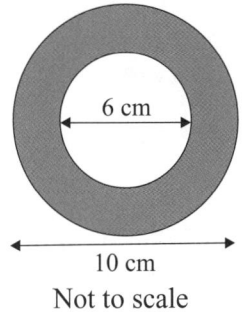

Calculate the area of the shaded region of the letter "O".
Give your answer to 2 decimal places.

.... cm²
[Total 3 marks]

3 Consider a square and a triangle. The sides of the square are x cm long. The base length and height of the triangle are equal, and are twice as long as the sides of the square. The area of the triangle is 9 cm² larger than the area of the square.

Find the perimeter of the square.

.... cm
[Total 4 marks]

4 The diagram shows Lynn's garden design.

Lynn's garden will be rectangular, with a semicircular flowerbed at each end.
The rest of the space will be taken up by a lawn.

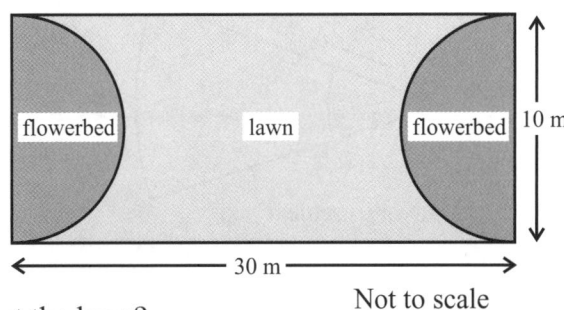

The grass seed that Lynn is planning to use comes in boxes that cost $7 each. Each box will cover 10 m².
How much will Lynn need to spend on grass seed to plant the lawn?

$ ....
[Total 3 marks]

Section Four — Geometry and Measures

5 An industrial rolling machine is made up of three identical cylinders of radius 9 cm.
The ends of the rollers are surrounded by a strip of metal, as illustrated in the diagram below.

Find the length of the metal strip,
giving your answer correct to 1 d.p.

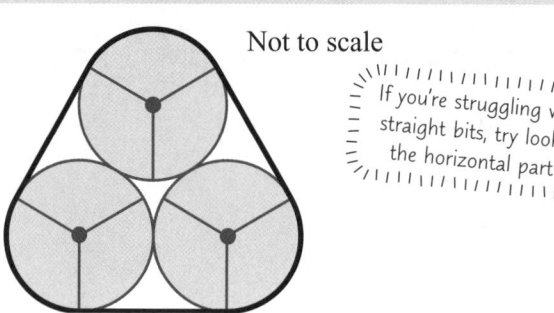

Not to scale

If you're struggling with the straight bits, try looking at the horizontal part first.

.... cm
[Total 4 marks]

6 Rectangle *B* is twice as long as rectangle *A*. They have the same width.
The two rectangles can be joined to make shape *C*, which has perimeter 28 cm.
They can be joined in a different way to make shape *D*, which has perimeter 34 cm.

Find the perimeters of rectangles *A* and *B*.

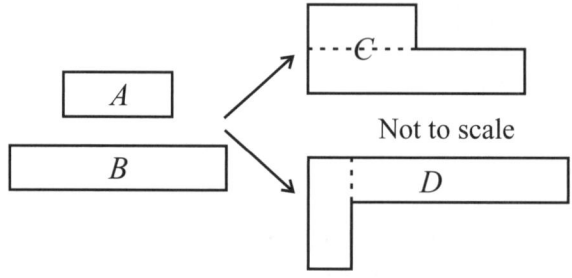

Not to scale

Perimeter of *A*: .... cm

Perimeter of *B*: .... cm
[Total 6 marks]

7 Look at the sector shown in the diagram below.

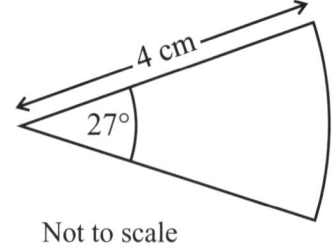

Not to scale

Find the perimeter and the area of the sector.
Give your answers to 3 significant figures.

Perimeter = ... cm

Area = ... cm²
[Total 5 marks]

Score: 27

Triangle Constructions

1 The diagram below is a sketch of triangle *ABC*.

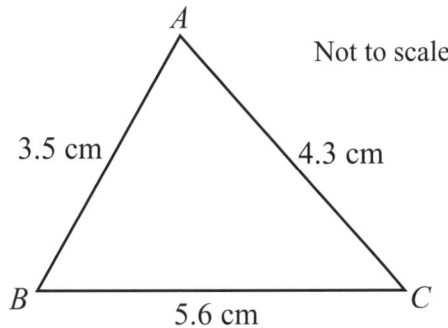

Not to scale

Use a ruler and compasses to make an accurate drawing of triangle *ABC* in the space below.
You must show all your construction lines.

Start by drawing one of the sides the correct length using your ruler.

[Total 2 marks]

2 *EFG* is an isosceles triangle. Sides *EG* and *FG* are both 4.5 cm long.

Side *EF* has been drawn below. Complete the construction of triangle *EFG* by drawing sides *EG* and *FG*.

[Total 2 marks]

Score:

4

Section Four — Geometry and Measures

3D Shapes — Surface Area and Nets

1 Juno wants to build a triangular prism out of cardboard.

a) Circle the net below that makes a triangular prism.

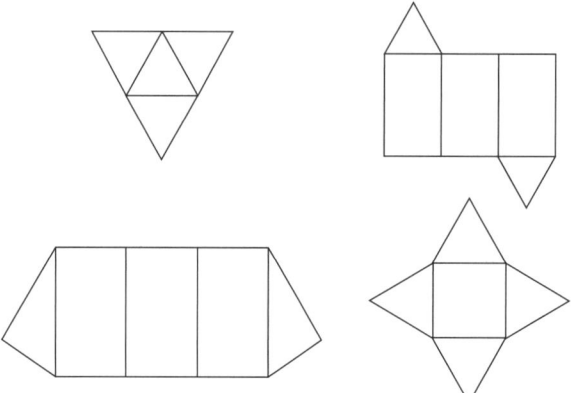

[1]

b) How many planes of symmetry does an equilateral triangular prism have?

....
[1]
[Total 2 marks]

2 A tower is made of 10 identical cubes, each with a side length of 3 m, stacked on top of each other, as shown.

Calculate the total surface area of the tower.

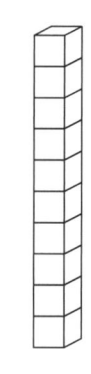

.... m²
[Total 3 marks]

3 Surface area of a sphere = $4\pi r^2$

Find the surface area of the sphere on the right.
Give your answer to 2 decimal places.

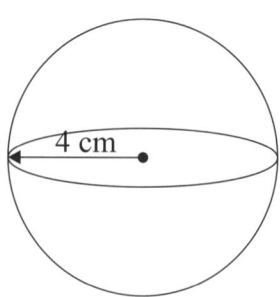

.... cm²
[Total 2 marks]

4 A regular triangle-based pyramid has four faces. Each face is an equilateral triangle. The pyramid can be made from the net shown.

Calculate the surface area of this pyramid.
Give your answer correct to 2 d.p.

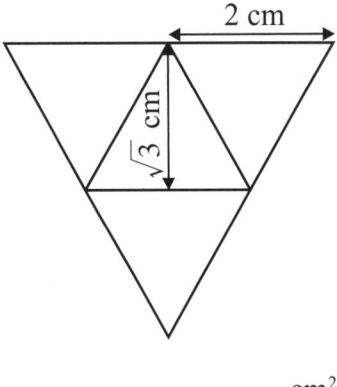

.... cm²
[Total 2 marks]

5 The diagram shows a regular octahedron. Each face of the octahedron is an equilateral triangle with base b and perpendicular height h.

a) Calculate the surface area of the octahedron when $b = 6.0$ cm and $h = 5.2$ cm.

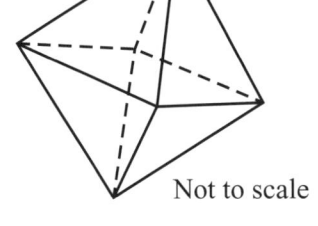

.... cm²
[2]

b) Show that a cylinder with radius b and height h has a greater surface area than the octahedron for any values of b and h.

[4]
[Total 6 marks]

6 The curved surface of a cone is made from the net shown.

a) The cone has a circular base.
Show that the radius of this circle is 4 cm.

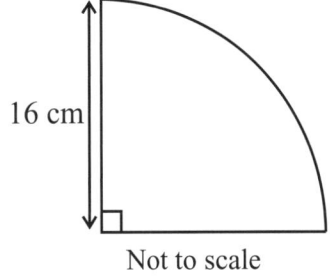

[2]

b) Calculate the total surface area of the cone. Give your answer to 2 decimal places.

Curved surface area of a cone = $\pi r l$

.... cm²
[3]
[Total 5 marks]

Score:
20

Section Four — Geometry and Measures

3D Shapes — Volume

1 Volume of sphere = $\frac{4}{3}\pi r^3$.

Find the volume of the sphere on the right.
Give your answer to 3 significant figures.

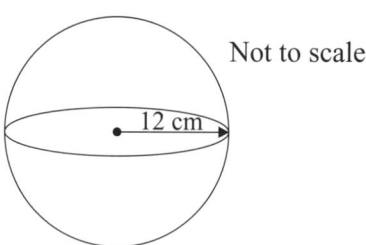
Not to scale
12 cm

.... cm³
[Total 2 marks]

2 The diagram below shows the dimensions of a triangular prism.

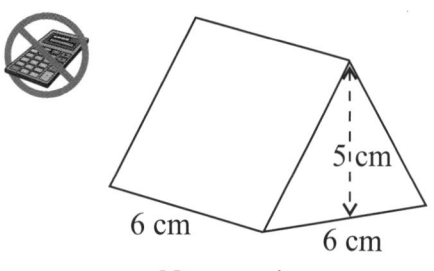

Calculate the volume of the triangular prism.

5 cm
6 cm   6 cm
Not to scale

.... cm³
[Total 2 marks]

3 A spherical ball has a volume of 478 cm³.

Find the surface area of the ball,
giving your answer correct to 1 d.p.

| Surface area of a sphere = $4\pi r^2$ |
| Volume of a sphere = $\frac{4}{3}\pi r^3$ |

.... cm²
[Total 4 marks]

4 The diagram on the right shows the dimensions of a grain hopper, which is made from a cuboid on top of a square-based pyramid.

Use the formula 'Volume of pyramid = $\frac{1}{3}$ × Base Area × Height' to calculate the volume of grain that the hopper can hold.

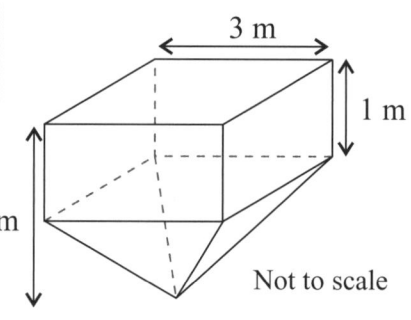
3 m
1 m
2.5 m
Not to scale

.... m³
[Total 3 marks]

5 The cone and sphere in the diagram below have the same volume.

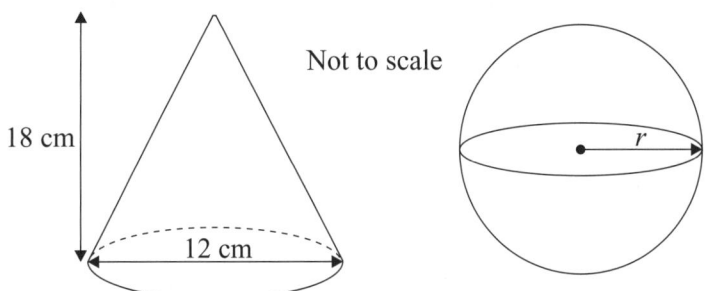

The cone has a vertical height of 18 cm and a base diameter of 12 cm.
Work out the radius, r, of the sphere. Give your answer to 3 significant figures.

.... cm
[Total 4 marks]

6 Water is flowing through the cylindrical pipe shown in the diagram below.
The radius of the pipe is 0.2 m, and the water comes halfway up the pipe.

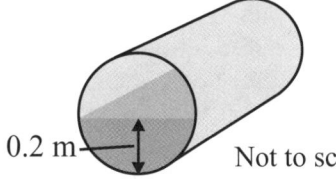

a) Find the cross-sectional area of the water in the pipe.

.... m^2
[1]

The water is flowing at a rate of 2520 litres per minute.

b) Find the speed of the water in m/s to 3 s.f.

Convert rate of flow to m^3/s:

2520 litres/minute = (2520 ÷ ....) litres/second = ... litres/second

= .... cm^3/s

= .... m^3/s

Speed = Rate of flow ÷ cross-sectional area of water

= .... m^3/s ÷ .... m^2

= .... m/s

.... m/s
[4]

[Total 5 marks]

Exam Practice Tip
Rate of flow questions like Q6 can be pretty confusing... Remember that rate of flow in m^3/s just means "volume moved per second". So dividing this by the cross-sectional area gives "distance moved per second" — and that's just speed. Don't rush into it though — make sure your units match up nicely first.

Score

20

Section Four — Geometry and Measures

Pythagoras' Theorem

1 The diagram shows a ramp placed against two steps.

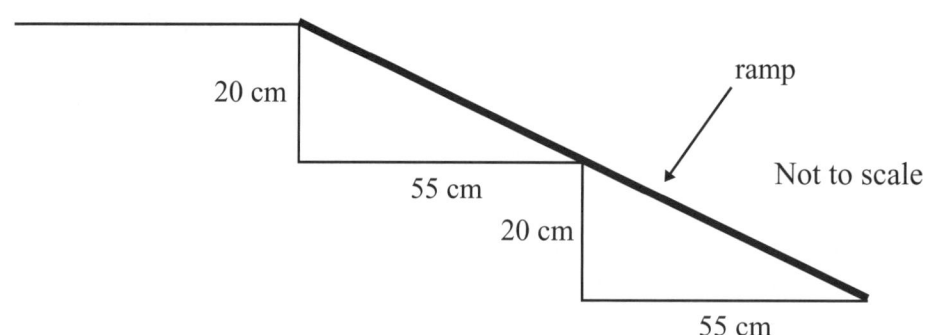

Calculate the length of the ramp.
Give your answer to 3 significant figures.

.... cm
[Total 3 marks]

2 A rectangle has a height of 3 cm and a diagonal length of 5 cm.

Calculate the area of the rectangle.

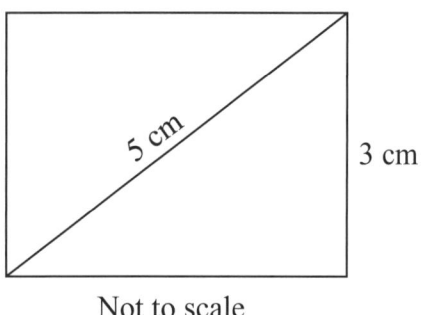

Not to scale

.... cm²
[Total 3 marks]

3 Point *A* has coordinates (2, –1). Point *B* has coordinates (8, 8).

Find the length of the line segment *AB*. Give your answer to 2 decimal places.

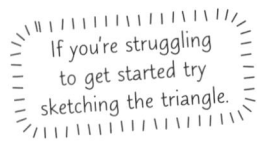
If you're struggling to get started try sketching the triangle.

....
[Total 3 marks]

Score:

9

Trigonometry — Sin, Cos and Tan

1 The diagram shows a right-angled triangle.

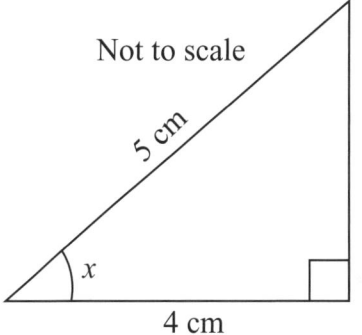

Not to scale

Work out the size of angle x. Give your answer to 1 decimal place.

$\cos x = \dfrac{\text{Adjacent}}{\text{..................................}}$

$\cos x = \dfrac{\text{............}}{\text{............}} = $ ....

$x = \cos^{-1}($ .... $) = $ ....

.... °

[Total 2 marks]

2 Quadrilateral *ABCD* is made up of two right-angled triangles, *ABC* and *ACD*.

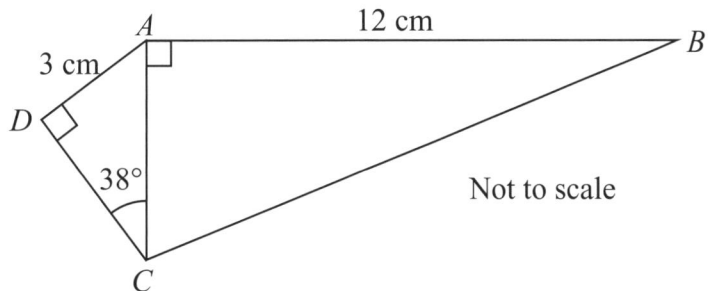

Not to scale

a) Find the lengths of the following sides. Give your answers to 2 decimal places.

i) *AC*

.... cm
[2]

ii) *DC*

.... cm
[2]

b) Use your answers to part a) to find the area of quadrilateral *ABCD*.
Give your answer to 2 significant figures.

.... cm²
[2]

[Total 6 marks]

3 The diagram shows a right-angled triangle.

 Find the exact length of the side marked y.

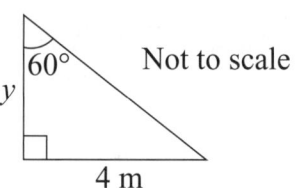

.... m
[Total 3 marks]

4 Sanjay is on a bridge, at point S, and looks down at a rat on the ground at point R.

Calculate the angle of depression of the rat from Sanjay. Give your answer to 1 decimal place.

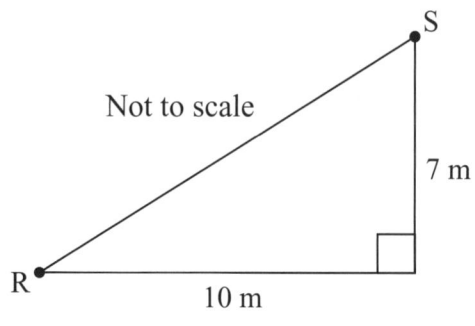

.... °
[Total 2 marks]

5 A swimmer sets off from the shore at point A and swims 200 m to point B, as shown in the diagram.

Find the shortest distance back to the shoreline from point B. Give your answer to 3 significant figures.

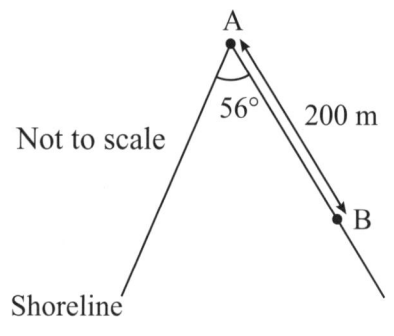

.... m
[Total 2 marks]

6 A regular hexagon is drawn such that all of its vertices are on the circumference of a circle of radius 8.5 cm.

The sum of interior angles in a polygon = (number of sides − 2) × 180°

Calculate the shortest distance from the centre of the circle to one of the edges of the hexagon. Give your answer to 2 decimal places.

.... cm
[Total 4 marks]

Exam Practice Tip

In an exam, it'll help if you start by labelling the sides of a right-angled triangle opposite (O), adjacent (A) and hypotenuse (H), so that you can see which trig formula you'll need to use. If you're working out an angle, make sure you check whether it's sensible — if you get an angle of 720° or 0.72°, it's probably wrong.

Score

19

The Sine and Cosine Rules

1 Look at triangle *PQR* below.

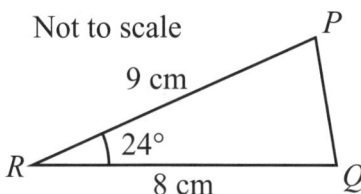

a) Calculate the length of side *PQ*.
Give your answer to 2 decimal places.

.... cm
[4]

b) Calculate the size of the acute angle *QPR*. Give your answer to 1 decimal place.

..........................°
[4]
[Total 8 marks]

2 In the triangle on the right, *AB* = 12 cm, *BC* = 19 cm and *AC* = 14 cm.

Calculate the area of the triangle.
Give your answer to 2 decimal places.

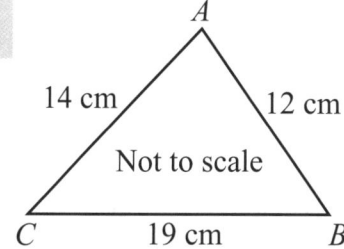

.... cm²
[Total 5 marks]

3 In the triangle below, *AB* = 17 cm, *AC* = 36 cm and angle *ACB* = 26°.
Angle *ABC* is obtuse.

Find the size of angle *ABC*. Give your answer to correct to 1 decimal place.

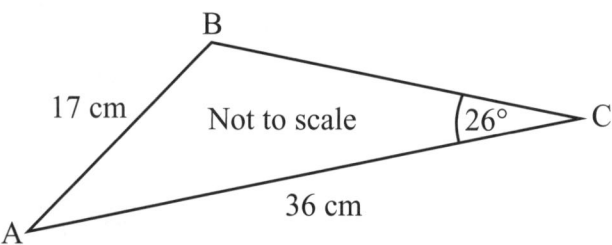

..........................°
[Total 4 marks]

Section Five — Pythagoras, Trigonometry and Vectors

4 *ABCD* is a quadrilateral.

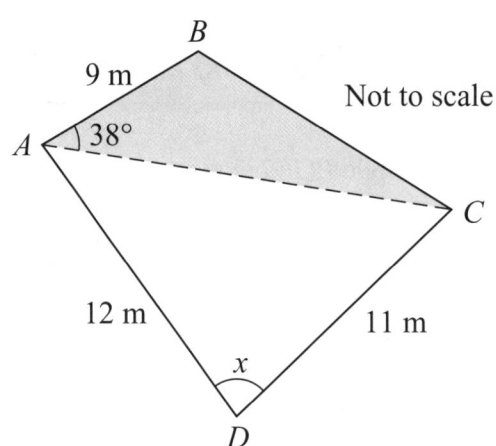

AB = 9 m.
AD = 12 m.
DC = 11 m.
Angle *BAC* = 38°.
Area of triangle *ABC* = 42 m².

Show that angle *x* is 82° to the nearest degree.

[Total 5 marks]

5 *ABCD* is a quadrilateral.

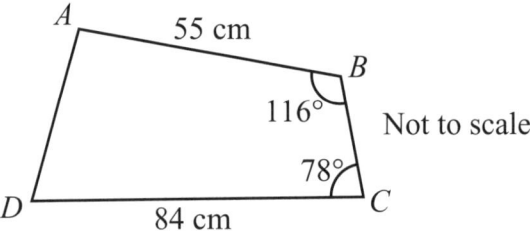

AB = 55 cm.
DC = 84 cm.
Angle *ABC* = 116°.
Angle *BCD* = 78°.

Given that *AC* = 93 cm, work out the area of *ABCD* to 3 significant figures.
Show clearly how you get your answer.

Add the line AC to the diagram to split the quadrilateral into two triangles.

.... cm²
[Total 6 marks]

Trig Graphs

1 Sketch the following trigonometric graphs, labelling the points where the graphs cross the *x*- and *y*-axes and any points where the graphs are undefined.

 a) $y = \cos x$ for $0° \leq x \leq 360°$

[2]

b) $y = \tan x$ for $0° \leq x \leq 360°$

[3]

[Total 5 marks]

2 The graph of $y = \sin x$ is shown for $0° \leq x \leq 360°$.

Solve the equation $4\sin x + 1 = 0$ for $0° \leq x \leq 360°$.
Give your answers correct to 1 decimal place.

You'll need to use a calculator to help you find the solutions. In this case, your calculator will give a negative value of x — add 360° to bring it into the right range, then use the graph to help you find the other solution.

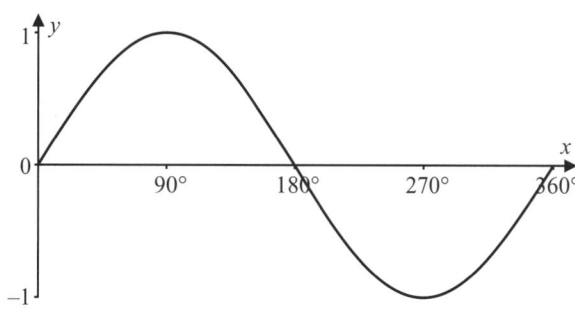

$x = $ .... and $x = $ ....

[Total 4 marks]

Exam Practice Tip

When you're solving equations involving trig functions it's important that you check your solutions are in the range specified in the question. Your calculator will only give you one solution and it might not be in the correct range — you'll have to use the graph of the trig function to find the right solutions.

Score

9

Section Five — Pythagoras, Trigonometry and Vectors

3D Pythagoras

1 The diagram on the right shows a cuboid ABCDEFGH. The cuboid has sides of length 6 cm, 4 cm and 3 cm.

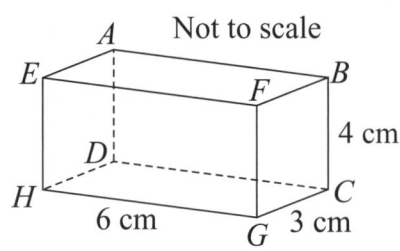

a) Calculate the length of the diagonal BH. Give your answer to 2 decimal places.

BH² = ...² + ...² + ...²

BH = √...............

BH = ...

.... cm
[3]

b) Another identical cuboid is joined to the end of this cuboid as shown on the right. Calculate the shortest distance from point M to point N. Give your answer to 2 decimal places.

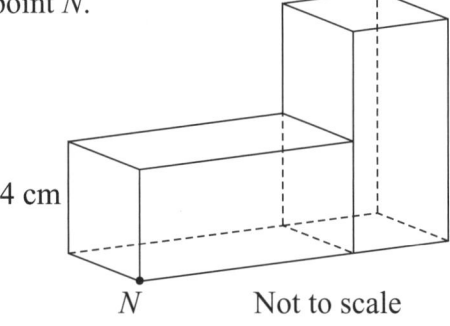

.... cm
[3]

[Total 6 marks]

2 The diagram shows a pyramid with a rectangular base. The vertex, V, of the pyramid is directly above X, the centre of the base ABCD.

VC = 8.9 cm, VX = 7.2 cm and BC = 4.2 cm.
Work out the length AB. Give your answer to 3 significant figures.

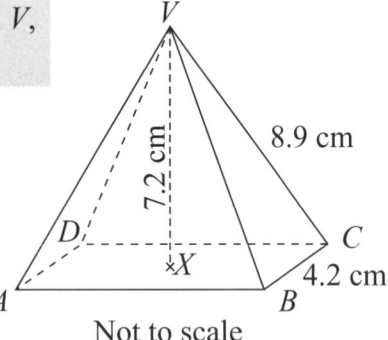

.... cm
[Total 4 marks]

Exam Practice Tip

If one of these questions pops up on your exam and you can't remember the formula for 3D Pythagoras then fear not — you can just break the problem down into a 2D Pythagoras question with a couple of steps. For example, in Q1 to find BH you could first find CH using triangle CGH then BH using triangle BCH.

Score: 10

3D Trigonometry

1 The cuboid *STUVWXYZ* has side lengths of 12 cm, 5 cm and 8 cm. *R* is the midpoint of *ST*.

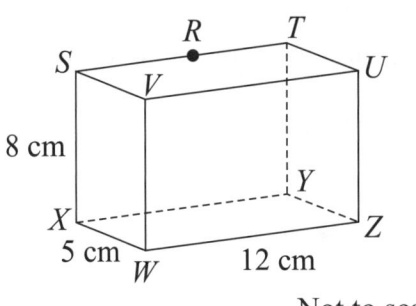

a) Find the angle that the diagonal *TW* makes with the base of the cuboid. Give your answer to 1 decimal place.

.... °
[4]

b) Show that the angle that *WR* makes with the top of the cuboid is 46° to the nearest degree.

[4]
[Total 8 marks]

2 The cone on the right has a slant height of 10 cm. The angle that the slant makes with the base of the cone is 72°.

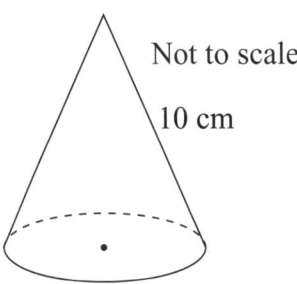

Calculate the total surface area of the cone.
Give your answer to 3 significant figures.

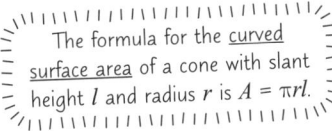

The formula for the curved surface area of a cone with slant height *l* and radius *r* is $A = \pi r l$.

.... cm²
[Total 4 marks]

Score: / 12

Vectors

1 **a**, **b** and **c** are column vectors, where $\mathbf{a} = \begin{pmatrix} -3 \\ 5 \end{pmatrix}$, $\mathbf{b} = \begin{pmatrix} 5 \\ 4 \end{pmatrix}$ and $\mathbf{c} = \begin{pmatrix} -4 \\ -6 \end{pmatrix}$.

Calculate:

a) **a** − **b**

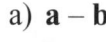

....
[1]

b) 4**b** − **c**

....
[1]

c) |2**a** + **b** + 3**c**|. Give your answer to 2 decimal places.

The straight lines mean you have to find the magnitude — use Pythagoras' theorem.

....
[3]
[Total 5 marks]

2 ABC is a triangle. $\overrightarrow{AB} = 2\mathbf{c}$ and $\overrightarrow{BC} = 2\mathbf{d}$. L is the midpoint of AC.

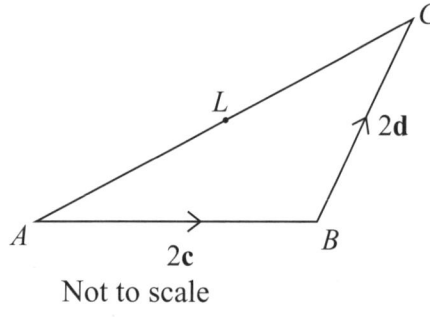

Not to scale

Write in terms of **c** and **d**:

a) $\overrightarrow{AL}$

$\frac{1}{2} \times \underline{} = \frac{1}{2} \times \underline{} + \frac{1}{2} \times \underline{}$

$= \underline{} + \underline{}$

....
[2]

b) $\overrightarrow{BL}$

....
[2]

c) Given that $\mathbf{d} = \begin{pmatrix} 2 \\ 5 \end{pmatrix}$, find $|\overrightarrow{BC}|$. Give your answer to 2 decimal places.

....
[3]
[Total 7 marks]

3   ABC is a triangle where $\vec{AB} = 3\mathbf{a} + \mathbf{b}$ and $\vec{CB} = -6\mathbf{a} + 4\mathbf{b}$. P is the midpoint of BC.

a) Write $\vec{AC}$ in terms of **a** and **b**.

....
[2]

b) Write $\vec{AP}$ in terms of **a** and **b**.

....
[2]

c) D is a point outside the triangle such that $\vec{CD} = -9\mathbf{a} - 3\mathbf{b}$.
Show that $\vec{CD}$ is parallel to $\vec{AB}$.

....

....
[2]

[Total 6 marks]

4   In the diagram, O is the origin.
$\vec{OA} = 2\mathbf{a}$ and $\vec{OB} = \mathbf{b}$. M is the midpoint of AB.

a) Find the position vector of M in terms of **a** and **b**.

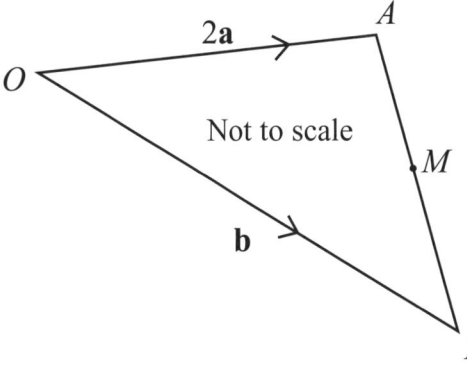

....
[2]

X is a point on AB such that AX : XB = 1 : 3.

b) Find the position vector of X in terms of **a** and **b**.

....
[3]

[Total 5 marks]

Exam Practice Tip
With vector questions, it's all about understanding the notation and knowing what you're trying to find — the actual calculations won't be that hard. Remember that vectors have a <u>direction</u> as well as a <u>length</u> — a common mistake is to think that $\vec{AB}$ and $\vec{BA}$ are equal, but actually $\vec{BA}$ is the reverse of $\vec{AB}$ so $\vec{AB} = -\vec{BA}$.

Score

23

Section Six — Probability and Statistics

Probability Basics

1 Sarah has stripy, spotty and plain socks in her drawer.
She picks out a sock from the drawer at random.
This table shows the probability of her picking a spotty sock.

Sock	Stripy	Spotty	Plain
Probability	$2x$	0.25	x

a) What is the probability that she picks a sock that is **not** spotty?

....
[1]

b) What is the probability of her picking a stripy sock?

Use your answer to part a) to find x.

....
[2]
[Total 3 marks]

2 Ami has a bag containing only strawberry and banana sweets in the ratio 2:5. She picks a sweet at random from the bag.

a) What is the probability that she picks a strawberry sweet from the bag?

....
[1]

b) Ami says, "I am exactly twice as likely to pick a banana sweet as a strawberry sweet".
Is Ami correct? Explain your answer.

....

....
[1]
[Total 2 marks]

3 There are p counters in a bag. n of the counters are blue and the rest are red. One counter is picked out at random.

Work out the probability that the counter picked is red.
Give your answer as a fraction in terms of p and n.

....
[Total 2 marks]

Score: ____ / 7

Section Six — Probability and Statistics

Finding Probabilities and Expected Frequency

1 Sumi has 3 pieces of homework, English (E), History (H) and Maths (M).
She has to do all 3 pieces tonight but she can do them in any order.

a) List all the different orders in which she could do her pieces of homework.

....

....
[2]

Sumi randomly chooses the order in which to do her pieces of homework.

b) Use your answer to part a) to find the probability that she does her Maths homework before her English homework. Give your answer as a fraction in its simplest form.

....
[1]

[Total 3 marks]

2 Katie decides to attend two new after-school activities. She can do one on Monday and one on Thursday. Below are lists of the activities she could do on these days.

Monday:	Hockey	Orchestra	Drama
Thursday:	Netball	Choir	Orienteering

a) List all nine possible combinations of two activities Katie could try in one week.

[2]

Katie randomly picks an activity to do on each day.

b) Use your answer to part a) to find:

  i) the probability that she does hockey on Monday and netball on Thursday,

....
[1]

  ii) the probability that she does drama on Monday.

....
[1]

c) Katie does 42 weeks of activities each year.
Estimate the number of times she will randomly choose to go to choir.

....
[2]

[Total 6 marks]

Section Six — Probability and Statistics

3 Alvar has a fair 6-sided dice and a set of five cards numbered 2, 4, 6, 8 and 10.
He rolls the dice and chooses a card at random.
Alvar adds the number on the dice to the number on the card to calculate his total score.

a) Complete the table below to show all of the possible scores.

		Cards				
		2	4	6	8	10
Dice	1					
	2					12
	3				11	13
	4			10	12	14
	5		9	11	13	15
	6	8	10	12	14	16

[2]

b) Find the probability that Alvar will score 12 or more.
Give your answer as a fraction in its simplest form.

....
[2]

c) Given that Alvar gets a score of 8, what is the probability that the card he chose was a 6?

....
[2]

[Total 6 marks]

4 The probability of a train arriving in Udderston on time is 0.64.
This year, Hester will get the train to Udderston 200 times and
Marquez will get the train to Udderston 300 times.

a) Estimate the number of times Hester will arrive in Udderston on time.

....
[1]

b) Estimate the number of times Marquez will not be on time when arriving in Udderston.

....
[2]

[Total 3 marks]

Exam Practice Tip
Expected frequency questions are sometimes linked to relative frequency questions (see p.107) — you might be given some experimental data and be asked to use relative frequencies to make estimates for a greater number of trials or a larger population. Just remember, Expected Frequency = Probability × Number of Trials.

Score

18

Section Six — Probability and Statistics

The AND/OR Rules

1 A biased 5-sided spinner is numbered 1-5.

The probability that the spinner will land on each of the numbers 1 to 5 is given in this table.

Number	1	2	3	4	5
Probability	0.3	0.15	0.2	0.25	0.1

a) What is the probability of the spinner landing on a 4 or a 5?

....
[2]

b) The spinner is spun twice. What is the probability that it will land on a 1 on the first spin and a 3 on the second spin?

....
[2]

[Total 4 marks]

2 Alisha and Anton are often late for dance class.
The probability that Alisha is late is 0.9. The probability that Anton is late is 0.8.

What is the probability that at least 1 of them is late to the next dance class?

The probability that at least one is late
= 1 − the probability that neither is late.

....
[Total 3 marks]

3 Reza is playing a game at a fair. The probability that he wins a prize is 0.3, independent of what has happened in previous games.

a) If he plays three games, what is the probability that he doesn't win a prize?

....
[2]

b) If he plays two games, what is the probability that he wins at least one prize?

....
[2]

c) Reza says, "If I play three games I have more than a 50% chance of winning exactly one prize". Is he correct? Explain your answer.

....

....
[3]
[Total 7 marks]

Score: ⬜
14

Tree Diagrams

1 Jo and Heather are meeting for coffee.
The probability that Jo will wear red trousers is $\frac{2}{5}$.
There is a one in four chance that Heather will wear red trousers.
The two events are independent.

a) Complete the tree diagram below.

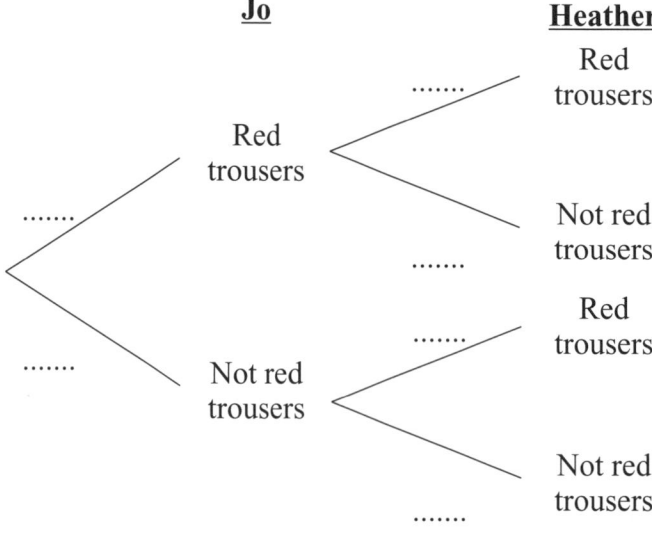

[2]

b) What is the probability that neither of them wear red trousers?

....
[2]
[Total 4 marks]

2 Paul and Jen play a game where they roll a fair dice. If it lands on a factor of 6 then Paul gets a point, otherwise Jen gets a point. The winner is the person who gets the most points.

a) If they roll the dice twice, what is the probability that it will be a draw?

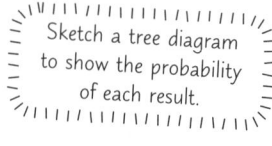
Sketch a tree diagram to show the probability of each result.

....
[3]

b) If they roll the dice three times, what is the probability that Paul wins?

....
[3]
[Total 6 marks]

3 The probability that Gemma has pasta for dinner depends on whether she had pasta the previous day. The probability that she will have pasta for dinner is 0.3 if she had it the previous day and 0.8 if she didn't have it the previous day.

a) Given that Gemma had pasta today, complete the tree diagram below.

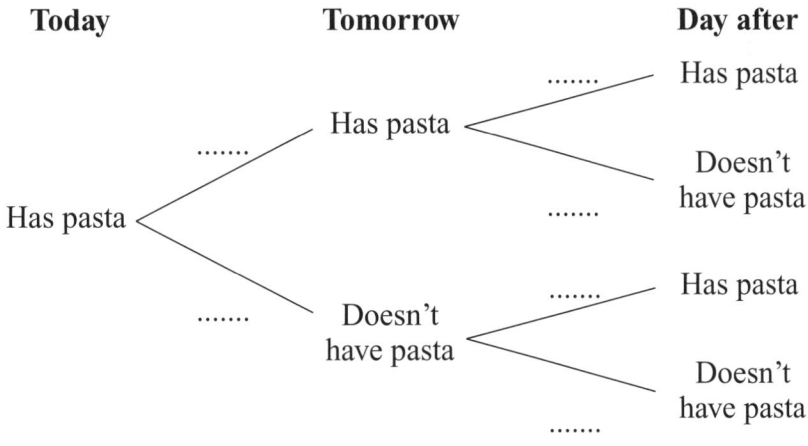

[2]

b) What is the probability of Gemma having pasta on exactly one of the next two days if she had pasta today?

P(pasta on 1 of the next 2 days) = P(..) + P(..)

= (.................. × ....) + (.................. × ....)

= .... + ... = ...

....
[2]

[Total 4 marks]

4 A bag of fruit contains 6 apples, 9 oranges and 5 pears.
Two pieces of fruit are chosen at random from the bag without replacement.

Find the probability that the two pieces of fruit are different.
Use the partially complete tree diagram to help.

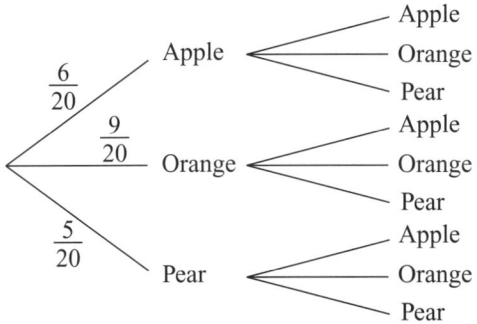

....
[Total 4 marks]

Exam Practice Tip
For questions with multiple events, it's helpful to use a tree diagram — even when you're not given one in the question (like Q2). They can help you keep track of all the possible outcomes so that you don't make any silly errors. Just make sure that you <u>multiply</u> along the branches to get the end probabilities.

Probability from Venn Diagrams

1 4000 people attended a baseball match. 1320 of them bought food, and 2560 bought a drink. 1080 people bought both.

a) Draw a Venn diagram to show the number of people who bought food, drinks, both or neither at the match.

[3]

b) Find the probability that a randomly selected member of the crowd bought food, but not a drink.

....
[1]

[Total 4 marks]

2 A cheese stall sells three different cheeses: Cheddar, Brie and Feta. One afternoon the stall had 100 customers. Each customer bought at least one cheese.

28 customers bought Brie. Of these, 12 customers also bought Feta.
43 customers bought Cheddar. Of these, 10 also bought Brie and 7 also bought Feta.
5 customers bought all three cheeses.

a) Draw a Venn diagram to show this information.

[3]

b) Find the probability that a randomly selected customer bought Cheddar or Feta.

....
[1]

c) Given that a customer bought Feta, find the probability that this customer also bought Brie.

....
[2]

[Total 6 marks]

Score: 10

Relative Frequency

1 John throws a ball at a target using his left and right hands. His results are shown in the table on the right.

	Left Hand	Right Hand
Throws	20	100
Hit target	12	30

a) Estimate the probability that John will hit the target with his next throw if he uses his left hand.

....
[2]

b) John uses his results to estimate the probabilities of him hitting the target using each hand. Explain which of his estimated probabilities will be more reliable.

....

....
[1]
[Total 3 marks]

2 Eimear has a bag containing a large number of counters. Each counter is numbered either 1, 2, 3, 4 or 5.

She selects one counter at random from the bag, makes a note of its number, and then puts it back in the bag. Eimear does this 100 times. Her results are shown in the table below.

Number on counter	1	2	3	4	5
Frequency	23	25	22	21	9
Relative Frequency					

a) Complete the table, giving the relative frequencies of each counter being selected.
[2]

b) Elvin says that he thinks that the bag contains the same number of counters with each number. Do you agree? Give a reason for your answer.

....
[1]

c) Using Eimear's results, estimate the probability of selecting an odd number when one counter is picked from the bag.

....
[2]
[Total 5 marks]

Score:

8

Section Six — Probability and Statistics

Organising Data

1 Two students, Cari and Martin, collected some data from their school.

Cari found out the star sign of each student in her class.

a) State whether Cari's data is quantitative or qualitative.

....
[1]

Martin found out the number of students late to school one day.
His data is quantitative.

b) State whether Martin's data is discrete or continuous.

....
[1]

[Total 2 marks]

2 Xin is investigating how many chocolate bars teenagers buy each week.
She is going to collect data by asking her teenage friends how many they buy.

a) Write down two problems with Xin using the question below in her survey.

> How many chocolate bars have you bought?
> 1 – 2   2 – 3   3 – 4

1 ....

2 ....
[2]

b) Xin also wants to ask how much her friends spend on chocolate each week. She designs a table to record her data, using inequalities.

Explain whether the inequalities she has used for the classes are suitable or not.

Money spent on chocolate	Tally	Frequency
£0 ≤ x < £2		
£2 ≤ x < £4		
£4 ≤ x		

....

....
[2]

c) Comment on whether she can use her results to draw conclusions about all teenagers.

....

....
[1]

[Total 5 marks]

Score: 7

Section Six — Probability and Statistics

Mean, Median, Mode and Range

1 One evening Preya makes 10 phone calls. When the bill comes it shows how long each call was, in minutes. The call lengths are listed below.

        10   12   25   3   37   13   12   18   41   33

a) Work out the median length of Preya's calls.

.... minutes
[2]

b) Calculate the mean phone call length. Give your answer to the nearest minute.

.... minutes
[2]
[Total 4 marks]

2 A bakery records the number of cookies it sells each day for ten days. The mean number is 17 and the median number is 15. The next day the bakery sells 18 cookies.

Comment on whether each of these statements is true or false. Explain your answers.

a) The mean number sold over all eleven days must be higher than 17.

....
[1]

b) The median number sold over all eleven days must be higher than 15.

....
[1]
[Total 2 marks]

3 Lee has 6 pygmy goats. Their weights, in kg, are listed below.

        32   23   31   28   36   26

a) Which three weights, from the list above, would have a range which is half the value of the median of the three weights? Write down the range and median with your answer.

.................., ..., ...

range = ..., median = ...
[2]

b) Two of the goats wander off and don't return. The mean weight of the herd is now 27.25 kg. Find the weights of the two goats who wandered off.

.... kg and .... kg
[3]
[Total 5 marks]

Exam Practice Tip
Questions on the mean might not be as straightforward as calculating its value from a list of numbers. You might have to figure out how to use the information you're given — e.g. if you know the mean and the number of values, you can easily find the total of all the values, which can let you find a missing value.

Score: 11

Frequency Tables

1 The table on the right shows the number of pets owned by the 29 pupils in a school class.

Number of pets	Frequency
0	8
1	3
2	5
3	8
4	4
5	1
6	0

a) Write down the range of this data.

.... *[1]*

b) Calculate the mean number of pets per pupil.

Add an extra column to the table to help you.

.... *[3]*

c) Find the median number of pets.

.... *[2]*

[Total 6 marks]

2 A quality control department checked the number of nails in 180 bags, each of which should contain 100 nails. The numbers of nails that they found in the bags are shown in the table below.

Number of nails per bag	97	98	99	100	101
Number of bags	6	20	44	108	2

a) Write down the modal number of nails in a bag.

.... *[1]*

b) What is the median number of nails in a bag?

.... *[2]*

c) Calculate the mean number of nails in a bag. Give your answer correct to 1 decimal place.

.... *[3]*

[Total 6 marks]

Score: 12

Section Six — Probability and Statistics

Grouped Frequency Tables

1 30 pupils in a class sat an exam in Science. The distribution of their marks is given in the table below.

Use the table to find:

Exam mark, x	Frequency
$10 < x \leq 20$	2
$20 < x \leq 30$	4
$30 < x \leq 40$	7
$40 < x \leq 50$	8
$50 < x \leq 60$	3
$60 < x \leq 70$	6

a) the modal class.

....
[1]

b) an estimate of the range.

....
[1]

c) an estimate of the mean.

Tip: add two columns to the table to help you.

....
[4]

[Total 6 marks]

2 The table shows the times it took 32 pupils at a school to run a 200 m sprint.

Time (t seconds)	Frequency
$22 < t \leq 26$	4
$26 < t \leq 30$	8
$30 < t \leq 34$	13
$34 < t \leq 38$	6
$38 < t \leq 42$	1

a) Calculate an estimate for the mean time.

.... seconds
[4]

b) What percentage of pupils got a time of more than 30 seconds?

.... %
[2]

c) Explain whether you could use the results in the table above to draw conclusions about how long it takes 16-year-old boys at the school to run 200 m.

....
[1]

[Total 7 marks]

Exam Practice Tip

With grouped frequency tables you can only make estimates of averages and spread because you don't know the exact value of each piece of data. If you have to estimate the mean in your exam, then you can just add columns to the table — you'll need one for the mid-interval value and one for frequency × mid-interval value.

Score 13

Simple Charts

1 The composite bar chart below shows the number of cups of different hot drinks sold in a cafe last Saturday and Sunday.

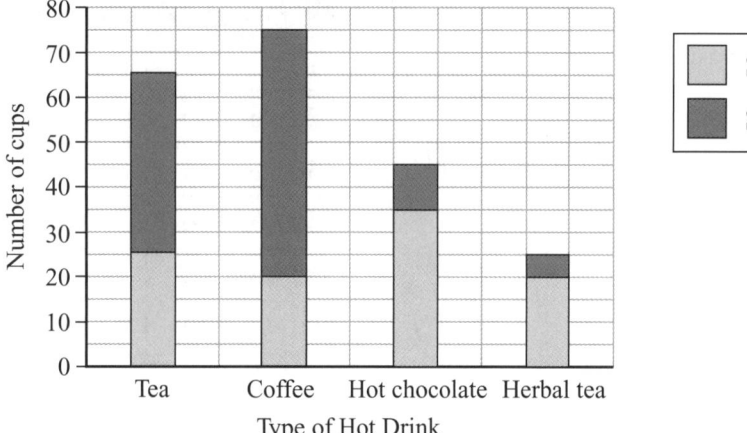

a) How many more cups of coffee were sold on Sunday than Saturday?

....
[1]

b) On which day were most hot drinks sold in total?

....
[2]

c) What fraction of the cups of herbal tea were sold on Saturday? Give your answer in its simplest form.

....
[1]
[Total 4 marks]

2 This pictogram shows the number of eggs laid by some chickens at a farm on Monday, Tuesday, Wednesday and Friday.

a) How many more eggs were laid on Wednesday than Tuesday?

....
[1]

b) 24 eggs were laid on Thursday. Show this information on the pictogram.

[1]

The farmer is going to use 40% of the eggs laid on Friday to make sponge cakes.

c) How many eggs is this?

....
[2]
[Total 4 marks]

3 This stem-and-leaf diagram shows the number of newspapers a shop sold on each day in June.

Key: 0 | 5 means 5 newspapers

```
0 | 0 0 0 2 2 5 8 9
1 | 1 1 1 2 3 4 5 7 9
2 | 0 0 5 6 6 7 7 7 8
3 | 0 3 5 9
```

a) On how many days did the shop not sell any newspapers?

.... days
[1]

b) On what fraction of the days did the shop sell more than 30 newspapers?
Give your answer in its simplest form.

....
[1]

c) What was the median number of newspapers sold by the shop in June?

....
[2]
[Total 4 marks]

4 This pictogram shows the number of jars of jam sold in a campsite shop in one month.

The key is missing from the pictogram.
The shop sold 100 jars of jam altogether.
How many jars of raspberry jam did it sell?

Strawberry Jam	🍓🍓🍓🍓
Blackberry Jam	🍓🍓◖
Raspberry Jam	🍓🍓🍓◖

There are .... symbols in total

So 1 symbol represents .... jars of jam

.... × .... = ... jars of raspberry jam

....
[Total 4 marks]

5 25 people were asked how many holidays they went on last year. The diagram below shows the results.

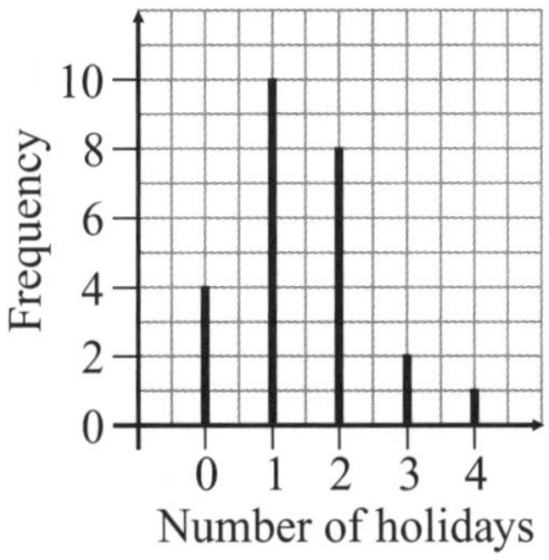

a) Find the mean number of holidays.

....
[3]

b) Find the median number of holidays.

Imagine the data in a list — 0, 0, 0, 0, 1... Find the position of the median and count up through the bars till you get there.

....
[2]
[Total 5 marks]

Score: ☐
21

Section Six — Probability and Statistics

Pie Charts

1 A survey was carried out in a cinema to find out which flavour of popcorn people bought. The results are in the table below.

a) Draw and label a pie chart to represent the information.

Type of popcorn	Number sold
Plain	12
Salted	18
Sugared	9
Toffee	21

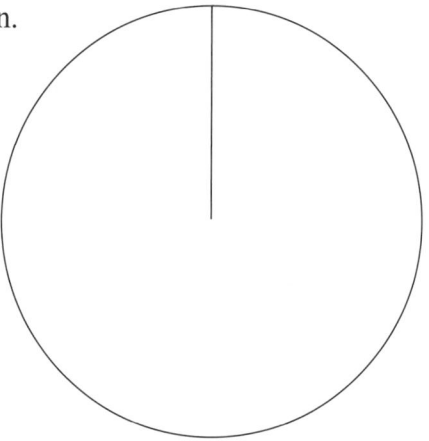

[4]

Another survey was carried out to find out which flavour of ice cream people bought. The results are shown in the pie chart below.

Chris compares the two pie charts and says,

"The results show that more people chose strawberry ice cream than toffee popcorn."

b) Explain whether or not Chris is right.

....
....
....
....

[1]

[Total 5 marks]

2 80 students were asked for their favourite type of soup. The pie chart below shows the results.

How many students chose leek & potato soup?

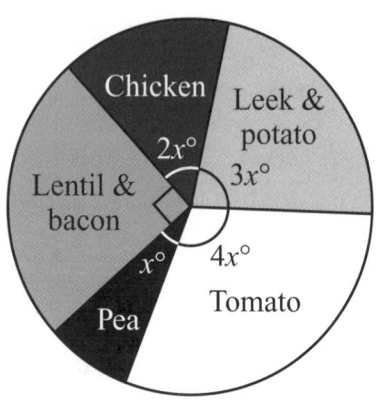

....

[Total 4 marks]

Score:

Section Six — Probability and Statistics

Scatter Diagrams

1 15 pupils in a class study both Spanish and Italian.
Their end of year exam results are shown on the scatter diagram below.

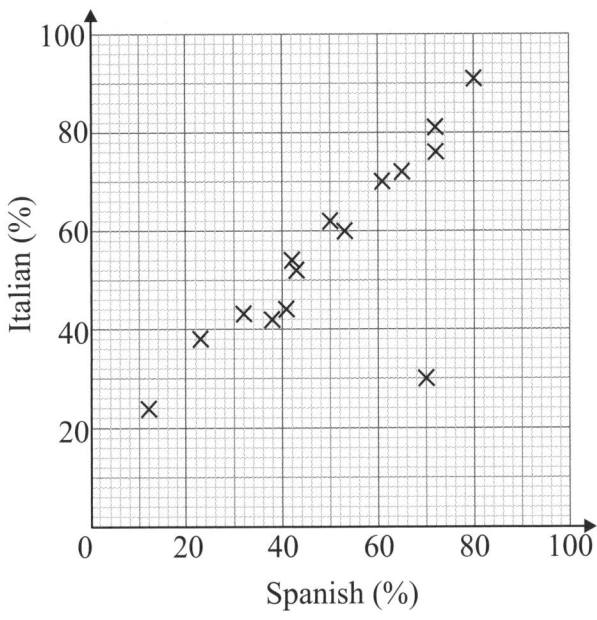

a) Circle the point that doesn't follow the trend.
[1]

b) Describe the strength and type of correlation shown by the points that do follow the trend.

....

....
[1]

c) Draw a line of best fit for the data.
[1]

[Total 3 marks]

2 A furniture company is looking at how effective their advertising is.
They are comparing how much they spent on advertising in random months with their total sales value for that month. This information is shown on the diagram below.

The table shows the amount spent on advertising and the value of sales for three more months.

Amount spent on advertising (thousands of $)	0.75	0.15	1.85
Sales (thousands of $)	105	60	170

a) Plot the information from the table on the scatter graph.
[1]

b) Describe the relationship between the amount spent on advertising and the value of sales.

....

....
[1]

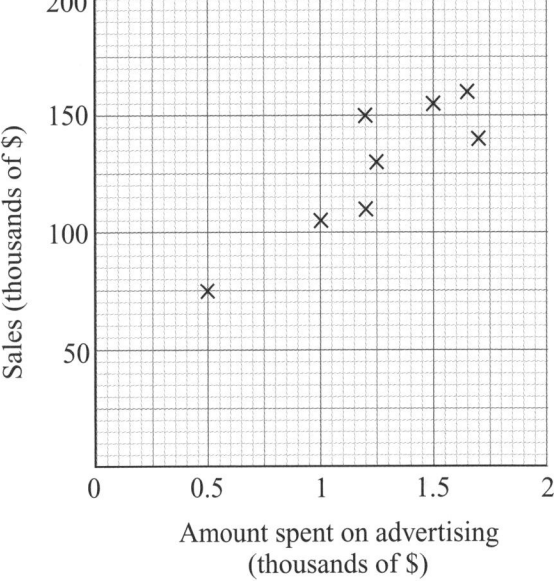

c) Use a line of best fit to estimate the amount spent on advertising in a month where the total value of sales was $125 000.

$ ....
[2]

[Total 4 marks]

Score:

7

Section Six — Probability and Statistics

Interpreting and Comparing Data

1   Two groups of people were surveyed about their favourite flavours of crisps. The results are shown below.

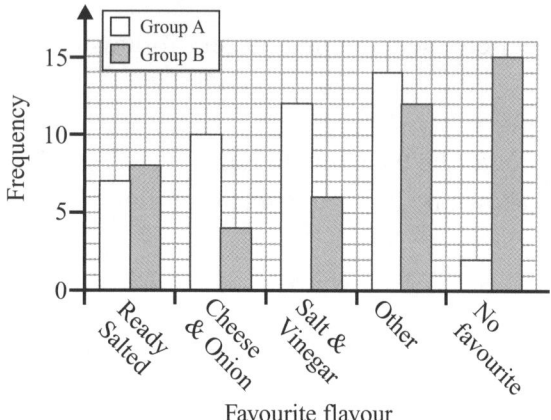

Decide whether each of the following statements is correct, and justify your answer.

a) "People in Group A are more likely to have a favourite flavour than people in Group B."

....

....

....

[1]

b) "There are no similarities between the preferences of the two groups."

....

....

[1]

[Total 2 marks]

2   A report on the levels of pollution caused by vehicles contains the graph below, which shows the number of flights per day from a particular airport over time.

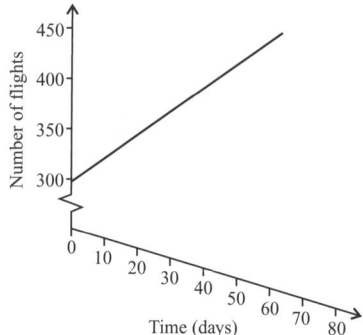

Give **two** reasons why this graph could be misleading.

....

....

....

....

[Total 2 marks]

3   A manufacturer tested the lifetimes (to the nearest 10 hours) of a type of light bulb so that they could confidently say how long they lasted. The results for eight such light bulbs are as follows:

2090	2010	2010	2550	90	2620	2800	2550

A customer asks the manufacturer how long, on average, this type of light bulb lasts. Which type of average would be appropriate to use? Give a reason for your answer.

....

....

....

[Total 2 marks]

Section Six — Probability and Statistics

4 The stem-and-leaf diagram below shows the runs scored by two cricketers in 15 matches.

```
        Cricketer A        Cricketer B
         9 0 0  | 0 | 5
         8 5 1  | 1 | 7 8
         5 5 3  | 2 | 0 6 8
         8 7 2  | 3 | 3 3 3 8 9
         3 1 0  | 4 | 2 3 6 7
```

3 | 4 = 43 runs   4 | 7 = 47 runs

E.g. You can compare the shapes of the two distributions, or refer to specific values from the diagram.

a) Compare the distributions of the runs scored by the two cricketers in context.

....

....
[2]

b) Which cricketer generally scored more runs? Give a reason for your answer.

....
[1]

[Total 3 marks]

5 A rugby club coach records the number of practice sessions each member attended in one year.

31   43   48   41   49   32   36   4   29   37   46   41   38

Would the range or interquartile range be more useful in analysing this data? Explain your answer.

....

....
[Total 2 marks]

6 Liz sells earrings. The prices in pounds of 15 pairs of earrings are given below.

6   3   4   8   10   11   5   7   4   12   8   9   5   7   11

a) Find the lower and upper quartiles of the prices above.

Start by writing the prices in ascending order.

Lower quartile = ...   Upper quartile = ...
[2]

b) Liz reduces all her prices by 50p. Will the interquartile range of the new prices be less than, greater than or the same as the interquartile range of the old prices? Give a reason for your answer.

....

....
[1]

[Total 3 marks]

Exam Practice Tip

In the exam, you could get asked to interpret or compare data from any types of table, chart or graph covered in this section. It's important that you are able to read the key information (averages and ranges) from them as well as identifying general trends. Just watch out for data presented in a misleading format.

Score

14

Histograms

1 A group of pupils were each given the same puzzle. The table below gives some information about how long it took the pupils to solve the puzzle.

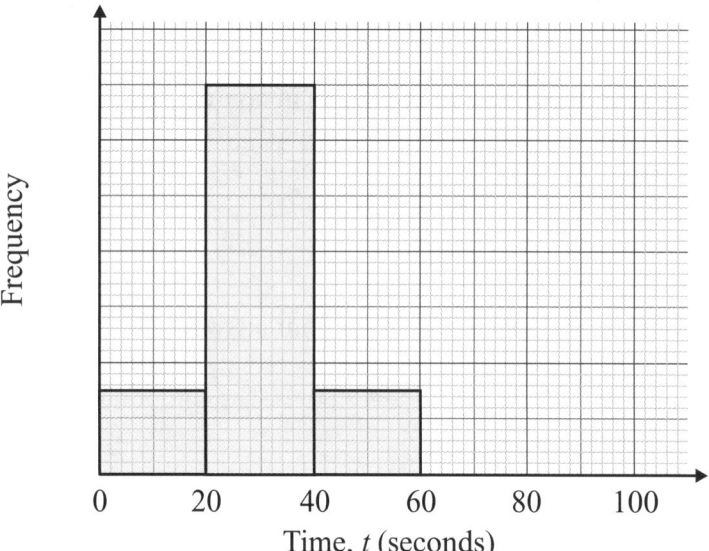

Time, t (s)	Frequency
$0 < t \leq 20$	15
$20 < t \leq 40$	
$40 < t \leq 60$	15
$60 < t \leq 80$	30
$80 < t \leq 100$	10

Don't forget to fill in the scale on the frequency axis.

Fill in the missing entry from the table and complete the histogram.

[Total 4 marks]

2 The histogram shows the amount of time (in minutes) that 270 children spent watching television one evening.

A large sample of adults were asked how long they watched television for on the same evening. The mean time for the adults was 102 minutes.

Does the data shown support the hypothesis that, on average, adults watched more television than children on this particular evening?

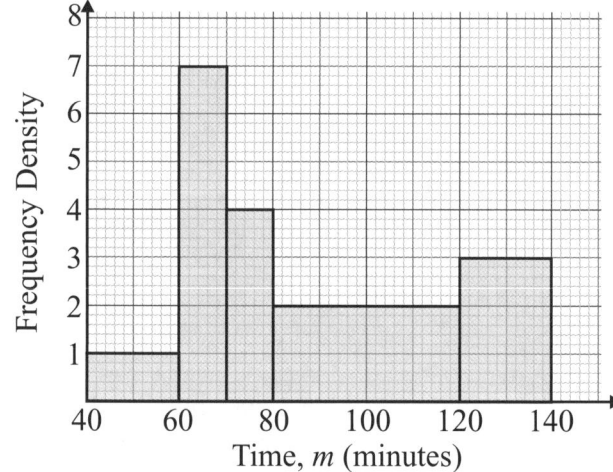

....

....

....

Make sure you show calculations to support your conclusion.

[Total 4 marks]

3 The histogram shows information about the weights, w kg, of 100 newborn lambs.

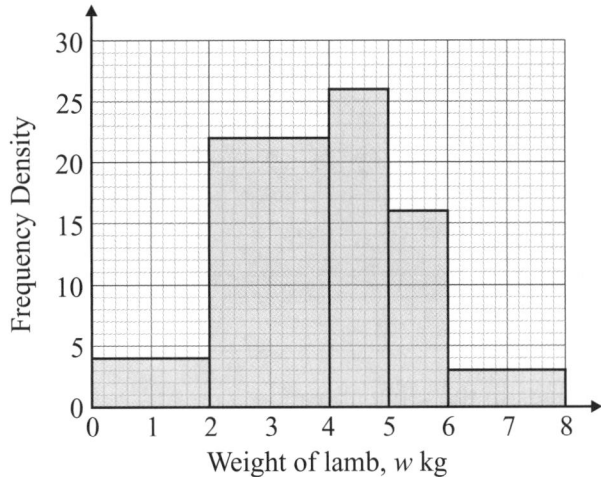

a) Calculate an estimate of the percentage of lambs weighing more than 3.5 kg.

.... %
[3]

b) This table shows information about the weights of the newborn lambs at a different farm.

Weight, w kg	$0 < w \leq 2$	$2 < w \leq 4$	$4 < w \leq 5$	$5 < w \leq 6$	$6 < w \leq 8$
Frequency	4	28	30	28	10

Draw a histogram on the grid to show this data.

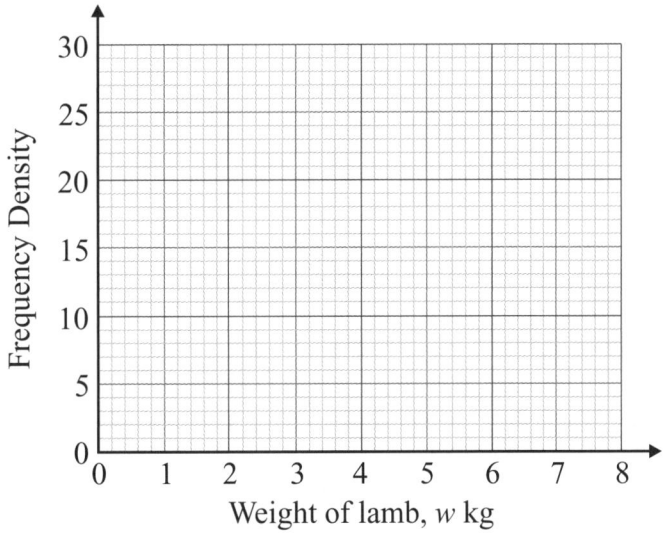

[4]

c) Compare the weights of newborn lambs for the two farms.

This question is only worth 1 mark, so you don't need to do any complicated calculations.

....

....
[1]
[Total 8 marks]

Exam Practice Tip

When the intervals in your histogram are unequal, you need to use frequency density rather than frequency on the vertical axis. For these histograms, the area of the bar is proportional to the frequency, so shorter bars that are really wide can sometimes represent a higher frequency than taller bars that are narrower.

Score

16

Cumulative Frequency

1   The times taken (*t* minutes) by 100 different kettles to boil a litre of water are shown in the table.

Time taken (*t* minutes)	$1.5 \leq t < 2$	$2 \leq t < 2.5$	$2.5 \leq t < 3$	$3 \leq t < 3.5$	$3.5 \leq t < 4$
Frequency	10	36	34	14	
Cumulative Frequency					

a)   Fill in all the missing values in the table above.

[2]

b)   Complete the cumulative frequency diagram below to show this data.

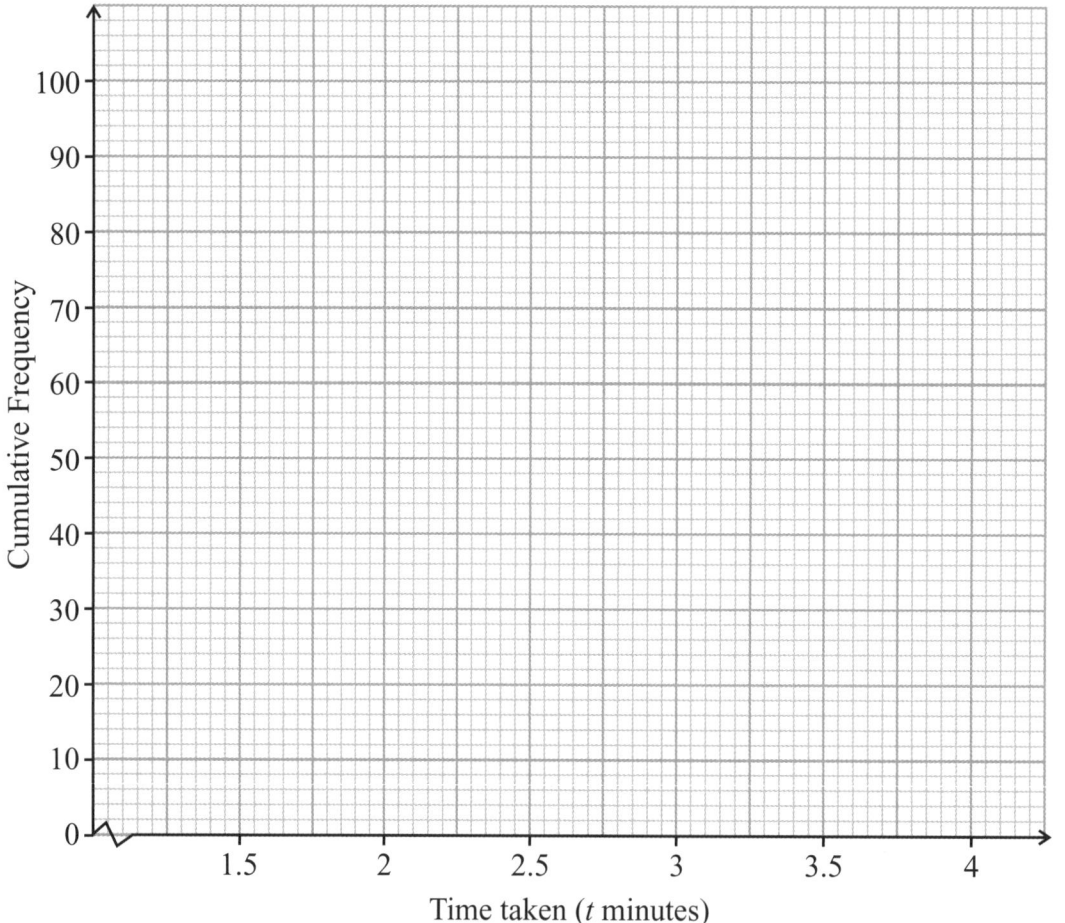

[2]

c)   Use your graph to find an estimate for the median time.

.... minutes

[2]

d)   Estimate the fraction of the kettles that took between 2 minutes and 3.4 minutes to boil the water. Give your answer in its simplest form.

....

[3]

[Total 9 marks]

2 The cumulative frequency diagram below shows information about the heights, in metres, of 80 trees in a forest.

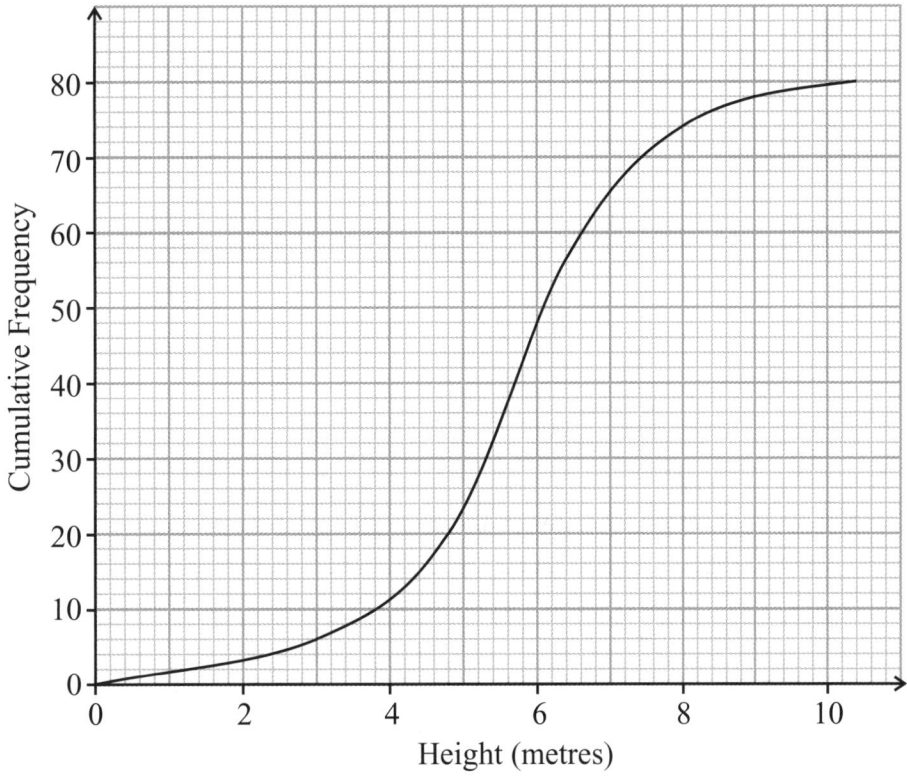

a) Use the cumulative frequency diagram to find an estimate of:

i) the interquartile range,

.... m
[2]

ii) the 65th percentile,

.... m
[2]

iii) the number of trees that are more than 8 metres tall.

....
[2]

b) Saul says, "More than 40% of the trees are between 4 m and 6 m tall."
Does the cumulative frequency diagram support this claim? Explain your answer.

....

....

....
[2]

[Total 8 marks]

Score: ☐

17

Section Six — Probability and Statistics

Mixed Questions

1 Natalie thinks of a natural number between 10 and 30.
Her number is not a prime number and when she squares her number, the final digit is 1.

What number did Natalie think of?

....
[Total 2 marks]

2 The Venn diagram contains elements of sets *A* and *B*.

$\xi = \{1, 2, 3, ..., 10\}$ $A = \{3, 4, 5, 6\}$ $B = \{x : x \text{ is a factor of } 12\}$

a) Complete the Venn diagram to show all the elements of each set.

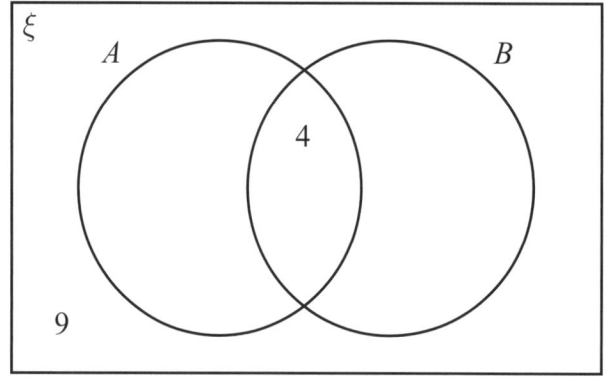

You need to give all the elements of each set, not just the number of elements in each one.

[2]

b) Write down n($A \cup B$).

....
[1]

c) A number from 1 to 10 is chosen at random.
Given that it is in set *A*, what is the probability that it is also in set *B*?

....
[1]
[Total 4 marks]

3 Work out $3\frac{3}{5} \div \frac{4}{7}$, giving your answer as a mixed number.

....
[Total 3 marks]

4 The *n*th term of a sequence is given by the formula $n^2 + 2n + 5$.

a) Fran says, "the 4th term in the sequence is a prime number."
Is Fran correct?

Show how you worked out your answer.

....

....
[2]

b) A different sequence begins 2, 5, 7, 12, 19, ...
Write down the next two terms in the sequence.

.... and ....
[2]

[Total 4 marks]

5 *AB* and *BC* are perpendicular lines which form the two mathematically similar triangles below.

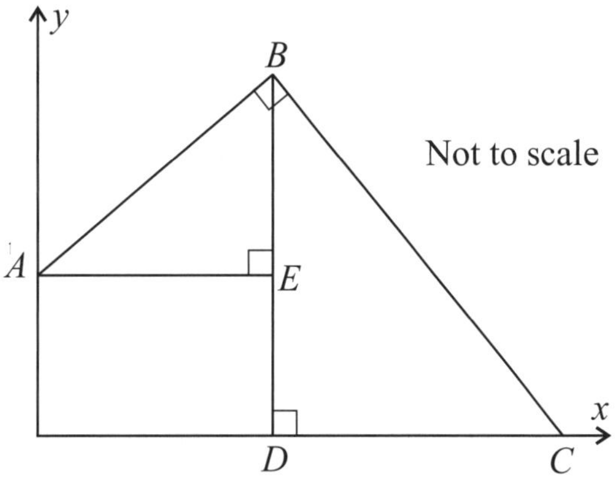

Not to scale

B has coordinates (12, 18).
C has coordinates (27, 0).
A is a point on the *y*-axis.
Angles *CBD* and *EAB* are equal.

a) Write the ratio *AE* : *BD* in its simplest form.

.... : ....
[2]

b) Work out the coordinates of *E*.

You'll need to use the fact that the triangles are similar.

(.............. , ...)
[3]

[Total 5 marks]

Mixed Questions

6 A factory produces brass screws and steel screws.

a) The factory produces 7×10^5 brass screws every day.
The probability of any brass screw being defective is 3.1×10^{-4}.

What is the expected number of defective brass screws every day?
Give your answer in standard form.

....
[2]

b) The probability of a steel screw being defective is $0.0\dot{4}$.

Write down this probability as a fraction. Show your working.

....
[2]
[Total 4 marks]

7 Solve the simultaneous equations to find values of *a* and *b*.

a) $3a + 2b = 17$
$2a + b = 10$

$a =$ ....
$b =$ ....
[3]

b) Hence, work out $a\begin{pmatrix}2\\1\end{pmatrix} - b\begin{pmatrix}3\\2\end{pmatrix}$

....
[2]
[Total 5 marks]

Mixed Questions

8 A company consists of 80 office assistants and 10 managers.
The pie chart shows how the 80 office assistants travel to work.

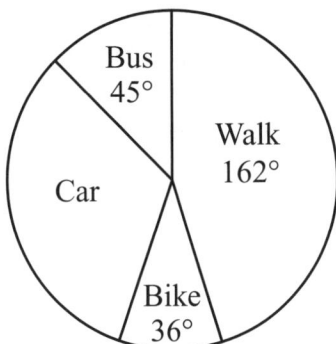

Office Assistants

a) How many office assistants travel to work by car?

.... [2]

All 10 of the managers travel by car.

b) What percentage of the people at the company travel by car?

.... %
[2]

Dave and Olivia are managers at the company and live on the same street.
The route between Dave and Olivia's street and the office is 28 kilometres.
Dave takes 42 minutes to drive to the office.
Olivia drives the same route. Her average speed is 10 kilometres per hour faster than Dave's.

c) How long does it take Olivia to drive from her house to the office?
   Give your answer rounded to the nearest minute.

.... minutes
[3]

d) Why is it important to your calculation that Olivia drives the same route as Dave?

....
....
[1]

[Total 8 marks]

9 Nigel runs a bakery.

a) Nigel wants to make 30 cupcakes using the recipe below.

Recipe for 12 cupcakes
140 grams butter
140 grams flour
132 grams sugar
2 eggs
1 tablespoon milk

i) How much sugar will Nigel need?

.... g
[2]

ii) Nigel measures 350.0 g of flour correct to the nearest 0.1 g.
Complete this statement about the mass of the flour, f grams.

.... ≤ f < ....
[2]

b) Ajay buys some packets of ginger biscuits and some packets of shortbread biscuits from Nigel's bakery.

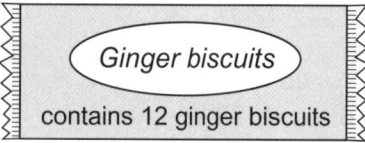
Ginger biscuits
contains 12 ginger biscuits

Shortbread biscuits
contains 10 shortbread biscuits

Ajay buys the same number of ginger biscuits and shortbread biscuits.
What is the smallest number of packets of shortbread biscuits that Ajay could have bought?

.... packets
[2]

c) Nigel currently pays $1.79 for a box of eggs.
If the price of the eggs increases by 0.4% each month, how much will a box of eggs cost Nigel in 2 years' time?

Careful here — this is an example of compound growth.

Give your answer to the nearest cent.

$ ....
[2]

[Total 8 marks]

10 The scatter graph shows the maximum power (in kW) and the maximum speed (in km/h) of a sample of cars.

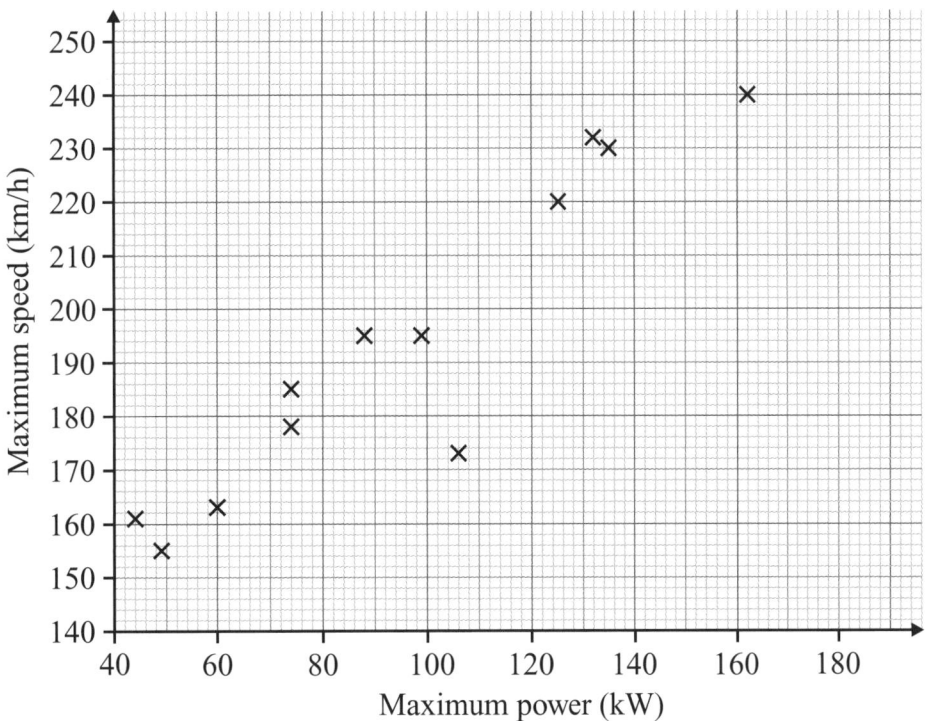

a) What percentage of the cars had a maximum speed less than 180 km/h?
Give your answer to 2 decimal places.

.... %
[2]

b) Ignoring the outlier, describe the correlation shown on the scatter graph.

.... correlation
[1]

c) A different car has a maximum power of 104 kW.
By drawing a suitable line on your scatter graph, estimate the maximum speed of this car.

.... km/h
[2]

d) What is the gradient of your line of best fit?
Give your answer to 1 decimal place.

....
[2]

e) Explain why it may not be reliable to use the scatter graph to estimate the maximum speed of a car with a maximum power of 190 kW.

....

....
[1]

[Total 8 marks]

Mixed Questions

11 Sides of two congruent nine-sided regular polygons form sides of quadrilateral *ABCD* and triangle *EFG*.

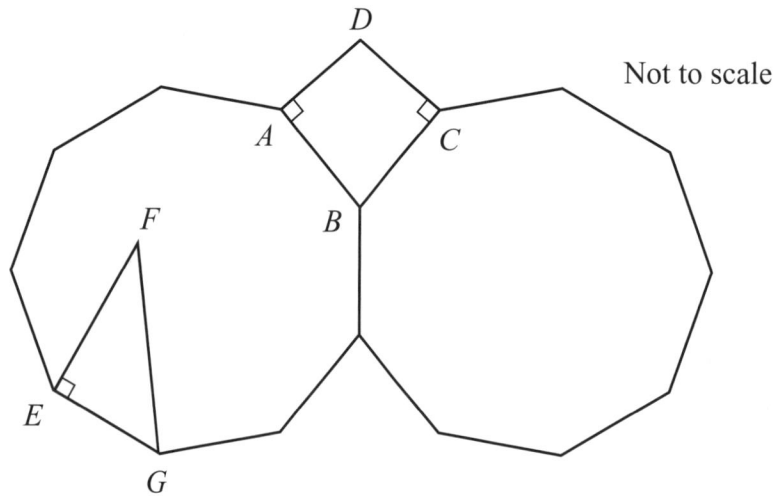

Not to scale

a) Angle *DAB* = angle *DCB* = 90°.
Calculate the size of angle *ADC*.

.... °
[3]

b) The ratio of the angles *EFG* : *EGF* : *FEG* is 2 : 3 : 5.

i) Find the angles in this triangle.

Angle *EFG* = ... °

Angle *EGF* = ... °

Angle *FEG* = ... °
[2]

ii) The hypotenuse of triangle *EFG* has a length of 15 cm and *AD* = 7.4 cm.
Calculate the total perimeter of the shape in the diagram above.

Give your answer correct to 1 decimal place.

Remember — SOHCAHTOA.

.... cm
[4]

[Total 9 marks]

12 Write $\frac{6}{\sqrt{3}} + \sqrt{27}$ in the form $k\sqrt{3}$.

....
[Total 3 marks]

13 Solve the following equations.

a) Solve $4^x = 256$

....
[1]

b) Solve $81^x = 3$

....
[2]
[Total 3 marks]

14 This cone is filled with water.

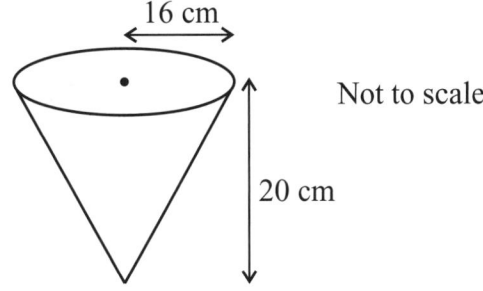

Not to scale

The radius of the cone is 16 cm to the nearest centimetre.
The height is 20 cm to the nearest centimetre.
Water leaks out of the bottom of the cone at a constant rate
of 0.39 litres per minute, to two significant figures.

Marion says, "The cone will definitely be empty after 15 minutes."
Is Marion correct? Explain your answer.
[The volume, V, of a cone with radius r and height h is $V = \frac{1}{3}\pi r^2 h$.]

....
[Total 5 marks]

15 Hannah and Tim both think of a number.
Hannah's number is negative. Tim's number is one more than Hannah's.

 They each take the reciprocal of their number. The sum of the reciprocals is $\frac{5}{6}$.
Use algebra to work out Hannah's original number.

....
[Total 5 marks]

16 The graphs below have equations $y = Ax^{-1}$, $y = Ax^n$ and $y = A^x$, where A is a positive integer.
Only the top-right quadrant of each graph is shown.

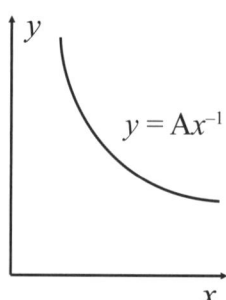 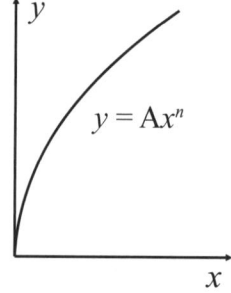 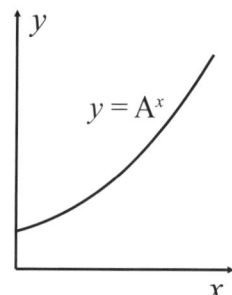

a) Circle the value of n.

-2 $\qquad$ $-\frac{1}{2}$ $\qquad$ $\frac{1}{2}$ $\qquad$ 2

[1]

b) Give the equation of the graph that has two lines of symmetry when all four quadrants are shown.

....
[1]
[Total 2 marks]

17 The line L passes through point A at (−2, −7) and point B at (3, 8).

a) Find the equation of line L between points A and B.

....
[3]

b) Find the coordinates of the midpoint of the line segment AB.

(.................. , ...)
[2]

c) Points x and y satisfy the following inequalities

$$x > 1 \qquad y > \frac{x}{2} \qquad x + 2y < 8$$

Show the region on the grid which satisfies these inequalities by shading the unwanted regions.

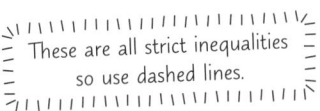
These are all strict inequalities so use dashed lines.

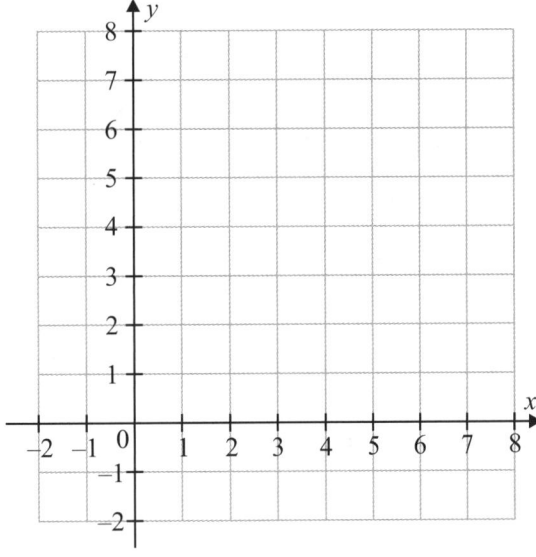

[4]
[Total 9 marks]

18 Write $\frac{10}{11}$ as a recurring decimal.

....
[Total 2 marks]

19 A, B and C are points on the circumference of a circle with centre O.

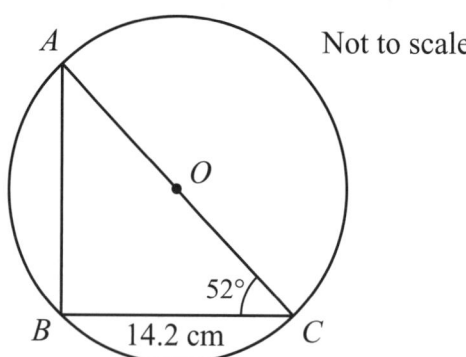

a) BC = 14.2 cm and angle ACB = 52°.
Calculate the circumference of the circle. Give your answer to 3 significant figures.

.... cm
[3]

b) D, E, F and G are points on the circumference of a different circle with centre O.

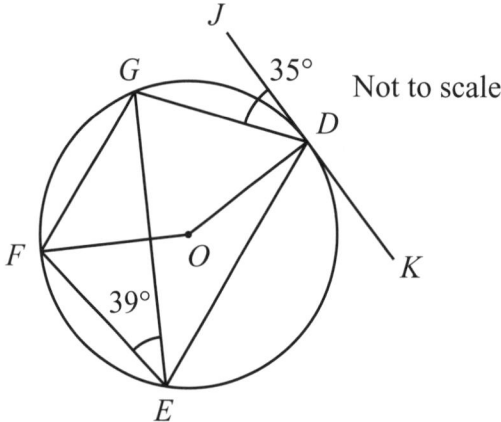

JK is a tangent to the circle at D. Angle JDG = 35° and angle GEF = 39°.
Work out the size of angle GFO.
Show all of your working, some of which may be on the diagram.

..............................°
[5]

[Total 8 marks]

20 This question is about the function $f(x) = \frac{9}{x^2} + 2x - 2$.

a) Complete the table below.

x	−4	−3	−2	−1	−0.5	0.5	1	2	3	4
f(x)	−9.4375		−3.75	5	33		9	4.25	5	6.5625

[2]

b) Use your table to draw the graph of $y = f(x)$ on the grid,
for values of x in the range $-4 \leq x \leq -0.5$ and $0.5 \leq x \leq 4$.

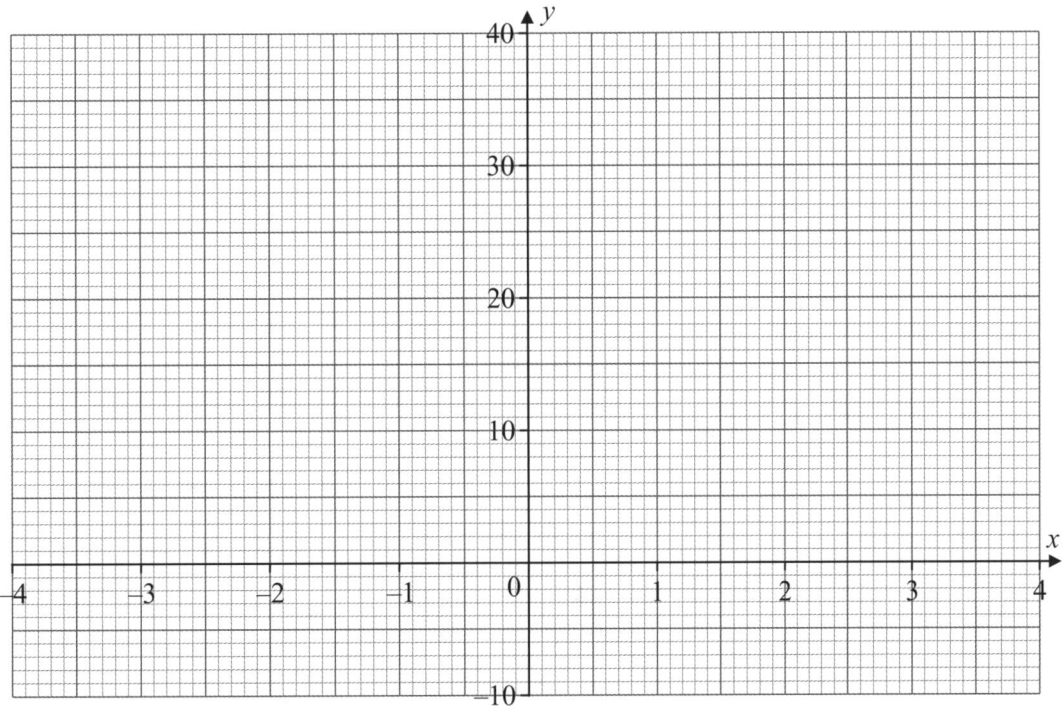

[5]

c) Use your graph to estimate the gradient of $f(x)$ when $x = 1.5$.

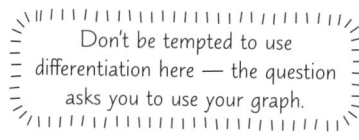

Don't be tempted to use differentiation here — the question asks you to use your graph.

....
[2]

d) By drawing a suitable straight line on the grid above, solve the equation $\frac{9}{x^2} + x - 10 = 0$.
Give your answers to 1 decimal place.

$x =$ .... and ....
[3]

[Total 12 marks]

21 The table shows some information about the ages of the adult members of a gym.

Age (A years)	$18 \leq A < 20$	$20 \leq A < 25$	$25 \leq A < 30$	$30 \leq A < 40$	$40 \leq A < 60$	$60 \leq A < 70$	$70 \leq A < 90$
Frequency	18	35	40	45	50	75	40

a) Draw a histogram for this data on the grid below.

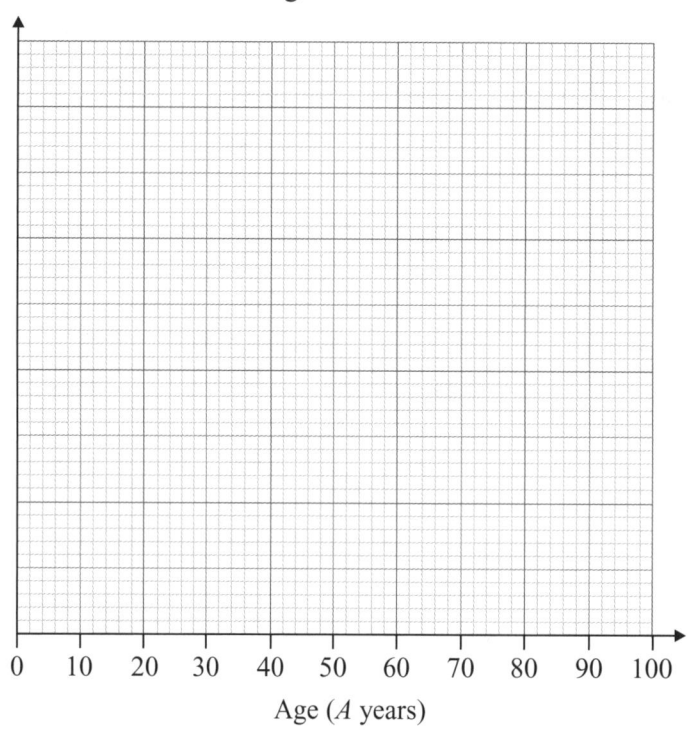

[5]

b) Estimate the mean age for the members of the gym.
Give your answer to the nearest whole year.

.... years
[4]

c) Explain why the mean does not give a very typical age for the members of this gym.

....

....
[1]

d) Two different adult gym members are chosen at random.
Find the probability that at least one of them is 30 years old or older.
Give your answer to 2 decimal places.

....
[3]
[Total 13 marks]

22 The functions f(x), g(x), h(x) and j(x) are shown below.

$$f(x) = 2x - 15 \qquad g(x) = x^2 + c, \text{ where } c \text{ is a constant} \qquad h(x) = 2x^2 \qquad j(x) = \frac{x^3}{3} + 2x^2 - 12x$$

a) Find f(–6)

....
[1]

b) Solve f(a) = 5

....
[1]

fg(4) = 25

c) Use this to find the value of c.

....
[2]

d) The domain of h(x) is {–2, 0, 2}. Find the range of h(x).

....
[1]

e) Find $\frac{dy}{dx}$ when y = j(x).

$\frac{dy}{dx}$ = ...
[2]

f) Find the coordinates of the turning points of the curve.

Remember — when you find the x-coordinates you'll have to substitute them into j(x) to get the y-values.

(................ , ...) and (................ , ...)
[4]

[Total 11 marks]

Exam Practice Tip
These mixed questions will give you a good idea about the variety of question types and topics you might have to tackle. In the actual exam you'll also be up against the clock — don't spend too long struggling with any one question, but move on and come back to it. And remember, always try to leave some time at the end to go back and check your answers.

Mixed Questions

Section One — Number

Page 3: Order of Operations

1. a) $4 \times 11 + 14 \div 2 \times 3 - 2 = 44 + 7 \times 3 - 2 = 44 + 21 - 2 = 63$
   [1 mark]
   b) $(3 \times 20 - 15) - (21 + 35 \div 7) = (60 - 15) - (21 + 5)$
   $= 45 - 26 = 19$ *[1 mark]*

2. $3 - 2 \times 7 + 9 \div 3 + 11 = -6$
   Try the brackets in different places
   until you get −6 on the right-hand side.
   $(7 + 9 \div 3) = 10$, so $3 - (2 \times 10) = -17$, then $-17 + 11 = -6$
   $3 - 2 \times (7 + 9 \div 3) + 11 = -6$ *[1 mark]*

3. $\dfrac{(13 \times 6 + 5)^2 + \sqrt{(6 \times 4) + (50 \times 20)}}{3^2} = 769$ *[1 mark]*

4. $\dfrac{197.8}{\sqrt{0.01} + 0.23} = 403.7575592688$ *[1 mark]*

   Your calculator might show a different number of decimal places, but don't worry — just write down what your display shows.

5. E.g. $3 = \sqrt{x^2 - 2y}$, so $9 = x^2 - 2y$
   Try different values of x and see what y-value each one gives:
   $x = 5$: $x^2 = 25$, so $2y = 25 - 9 = 16$, so $y = 8$
   $x = 7$: $x^2 = 49$, so $2y = 49 - 9 = 40$, so $y = 20$
   [2 marks available — 1 mark for each correct pair of x and y values]
   This equation works for any value of x that's an odd number and ≥ 5.

Page 4: Ordering by Size and Negative Numbers

1. $7 + 2 < 3 + 9 - 2$ *[1 mark]*
   $-4 + 8 + 2 \geq 3 + 3$ *[1 mark]*
   [2 marks available in total — as above]

2. −1.12, −0.61, −0.23, 0.35, 0.75, 1.06 *[1 mark]*

3. $1 \div (-5) = -\dfrac{1}{5} = -0.2$ *[1 mark]*

4. 0.035, 0.0355, 0.335, 0.35, 0.355, 0.503, 0.53 *[1 mark]*

5. $14 - -4 = 14 + 4 = 18$ °C *[1 mark]*

6. $(3 - -4) \times 5 = 7 \times 5 = 35$
   [2 marks available — 1 mark for (3 − −4) × 5, 1 mark for the correct answer]
   You might need to use trial and error for this one.

Page 5: Special Types of Number

1. $0.54^6 + \sqrt[3]{2.87} = 1.4459036... = 1.45$ (3 s.f.) *[1 mark]*

2. a) 18, −22, $\sqrt{16}$ (= 4) *[1 mark]*
   b) $\sqrt{3}$ and 5π *[1 mark]*

3. $\dfrac{\sqrt{6}}{4\sqrt{(10 - 2 \times 2)}} = \dfrac{\sqrt{6}}{4\sqrt{6}} = \dfrac{1}{4}$ — the expression is rational.
   [2 marks available — 1 mark for $\dfrac{\sqrt{6}}{4\sqrt{6}}$, 1 mark for correctly stating that $\dfrac{1}{4}$ is a rational number]

4. a) $\sqrt{270} = 16.4...$, $\sqrt{300} = 17.3...$,
      so square number = $17^2 = 289$ *[1 mark]*
   b) $\sqrt[3]{320} = 6.8...$, $\sqrt[3]{360} = 7.1...$,
      so cube number = $7^3 = 343$ *[1 mark]*

Page 6: Multiples and Factors

1. a) 72 *[1 mark]*
   b) 64 *[1 mark]*

2. a) 1, 2, 4, 7, 14, 28 *[1 mark]*
   b) 56, 64 *[1 mark]*

3. E.g. 4 (even) has three factors (1, 2 and 4).
   81 (odd) has five factors (1, 3, 9, 27 and 81).
   [2 marks available — 1 mark for each correct example of odd and even square numbers with a suitable number of factors]
   These aren't the only square numbers that would work here — any pair where the odd number doesn't have fewer factors than the even number would get you the marks.

4. The common multiples of 6 and 7 are:
   42, 84, 126, 168, 210, 252, ... *[1 mark]*
   42 and 84 are both common factors of 252 and 420 *[1 mark]*
   $x > 50$, so $x = 84$ *[1 mark]*
   [3 marks available in total — as above]

Page 7: Prime Numbers and Prime Factors

1. a) 7 or 11 *[1 mark]*
   b) E.g. 7 and 12 *[1 mark]*
   There are a few correct answers here — 1 and 12 or 11 and 12 are also correct.
   c) 7 *[1 mark]*
   $7 \times 2 \times 2 \times 3 = 84$.

2. Jack is incorrect as there are four prime numbers (101, 103, 107 and 109) between 100 and 110.
   [2 marks available — 1 mark for stating that Jack is incorrect, 1 mark for providing evidence]
   Writing any prime number between 100 and 110 is enough evidence.

3. E.g. 37 ($3 + 7 = 10$, which is 1 more than 9, a square number)
   [2 marks available — 2 marks for a correct answer, otherwise 1 mark for either a two-digit prime number or a number whose digits add up to 1 more than a square number]

4.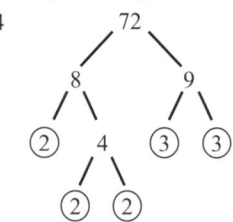

   $72 = 2^3 \times 3^2$ or $2 \times 2 \times 2 \times 3 \times 3$
   [2 marks available — 2 marks for a correct product of only prime factors, otherwise 1 mark for identifying the prime factors 2 and 3, or for a correct product containing non-prime factors]

Page 8: LCM and HCF

1. Factors of 12 = 1, 2, 3, 4, 6, 12
   Factors of 32 = 1, 2, 4, 8, 16, 32
   HCF = 4 *[1 mark]*

2. a) LCM = $3^7 \times 7^3 \times 11^2$ *[1 mark]*
   b) HCF = $3^4 \times 11$ *[1 mark]*

3. Multiples of 15 are: 15, 30, 45, 60, 75, 90, 105, 120, ...
   Multiples of 24 are: 24, 48, 72, 96, 120, ...
   LCM = 120
   [2 marks available — 1 mark for a correct method to find LCM, 1 mark for LCM correct]

4. Multiples of 16 are: 16, 32, 48, 64, 80, 96, 112, 128, 144, 160, ...
   Multiples of 36 are: 36, 72, 108, 144, 180, ...
   The LCM of 16 and 36 is 144, which is the minimum number of each item she needs.
   The minimum number of packs of jars she needs is
   $144 \div 16 = 9$ packs
   The minimum number of packs of labels she needs is
   $144 \div 36 = 4$ packs
   [3 marks available — 1 mark for a correct method to find LCM, 1 mark for LCM correct, 1 mark for both correct number of packs]

Pages 9-10: Fractions

1. E.g. $\dfrac{32}{42}$ *[1 mark]*
   Your answer can be any fraction where the top and bottom have been multiplied by the same integer > 1.

2. $\dfrac{40}{15} = 2\dfrac{10}{15} = 2\dfrac{2}{3}$
   [2 marks available — 1 mark for the correct improper fraction, 1 mark for the correct final answer]

3 a) $3\frac{1}{2} + 2\frac{3}{5} = \frac{7}{2} + \frac{13}{5} = \frac{35}{10} + \frac{26}{10} = \frac{35+26}{10} = \frac{61}{10} = 6\frac{1}{10}$

   [3 marks available — 1 mark for finding a correct common denominator, 1 mark for a correct addition using the common denominator, 1 mark for the correct answer as a mixed number]

   b) $3\frac{3}{4} - 2\frac{1}{3} = \frac{15}{4} - \frac{7}{3} = \frac{45}{12} - \frac{28}{12} = \frac{45-28}{12} = \frac{17}{12} = 1\frac{5}{12}$

   [3 marks available — 1 mark for finding a correct common denominator, 1 mark for a correct subtraction using the common denominator, 1 mark for the correct answer as a mixed number]

   If you've used a different method in Q3, but still shown your working, and ended up with the same final answer, then you still get full marks.

4 a) $1\frac{2}{3} \times \frac{9}{10} = \frac{5}{3} \times \frac{9}{10} = \frac{45}{30} = \frac{3}{2} = 1\frac{1}{2}$

   [3 marks available — 1 mark for converting the mixed number to an improper fraction, 1 mark for multiplying, 1 mark for the correct final answer]

   b) $3\frac{1}{2} \div 1\frac{2}{5} = \frac{7}{2} \div \frac{7}{5} = \frac{7}{2} \times \frac{5}{7} = \frac{35}{14} = \frac{5}{2} = 2\frac{1}{2}$

   [3 marks available — 1 mark for converting both mixed numbers to improper fractions, 1 mark for multiplying by the reciprocal, 1 mark for the correct final answer]

5 $\frac{2}{5} + \left(1 - \frac{6}{7}\right) = \frac{2}{5} + \frac{1}{7} = \frac{14}{35} + \frac{5}{35} = \frac{19}{35}$

   [3 marks available — 1 mark for finding the unshaded region of shape Y, 1 mark for writing over a common denominator and 1 mark for the correct answer]

6 $5 \times \frac{5}{6} = \frac{25}{6} = 4\frac{1}{6}$ *[1 mark]*, so they will need 5 pizzas *[1 mark]*.
   Cost = $(2 \times \$22) + \$12.50 = \$56.50$ *[1 mark]*
   [3 marks available in total — as above]

7 $17\frac{1}{2} \times \frac{1}{5} = \frac{35}{2} \times \frac{1}{5} = \frac{35}{10} = \frac{7}{2}$ *[1 mark]* tonnes of flour used to make cheese scones.
   Then $\frac{7}{2}$ out of 25 = $\frac{7}{2} \div 25 = \frac{7}{50}$ *[1 mark]*.
   [2 marks available in total — as above]

8 Shaded regions are $\frac{1}{4}$, $\frac{1}{4} \times \frac{1}{4} = \frac{1}{16}$ and $\frac{1}{4} \times \frac{1}{4} \times \frac{1}{4} = \frac{1}{64}$
   So total area shaded = $\frac{1}{4} + \frac{1}{16} + \frac{1}{64} = \frac{16}{64} + \frac{4}{64} + \frac{1}{64} = \frac{21}{64}$
   [3 marks available — 1 mark for working out the fraction for each shaded region, 1 mark for writing over a common denominator and 1 mark for correct answer]

Page 11: Fractions, Decimals and Percentages

1 a) $0.24 = \frac{24}{100} = \frac{6}{25}$ *[1 mark]*
   b) $0.06 \times 100 = 6\%$ *[1 mark]*

2 $65\% = 0.65$, $\frac{2}{3} = 0.666...$, $\frac{33}{50} = 0.66$
   So order is 0.065, 65%, $\frac{33}{50}$, $\frac{2}{3}$ *[1 mark]*

3 $\frac{1}{4} = 25\%$, so Jenny pays $100\% - 25\% - 20\% - 20\%$
   = $100\% - 65\% = 35\%$ *[1 mark]*
   $\$17.50 = 35\%$, so $1\% = \$17.50 \div 35 = \$0.50$ *[1 mark]*.
   The total bill was $\$0.50 \times 100 = \$50$ *[1 mark]*.
   [3 marks available in total — as above]

4 Let $r = 0.\dot{7}$, so $10r = 7.\dot{7}$
   $10r - r = 7.\dot{7} - 0.\dot{7}$ *[1 mark]*
   $9r = 7$, so $r = \frac{7}{9}$ *[1 mark]*
   [2 marks available in total — as above]

5 Convert to an equivalent fraction with all nines on the bottom:
   $\frac{7}{33} = \frac{21}{99}$ *[1 mark]*
   Then the number on the top tells you the recurring part, so $\frac{7}{33} = 0.\dot{2}\dot{1}$ *[1 mark]*
   [2 marks available in total — as above]

Pages 12-14: Percentages

1 For every 2 grapes there are 5 cherries so there are $\frac{2}{5} = 40\%$ as many grapes as cherries. *[1 mark]*

2 $\frac{72}{240} \times 100 = 30\%$ *[1 mark]*

3 $278 + 197 - 200 = 275$ books are left in the library.
   44% of 275 = $275 \times 0.44 = 121$ fiction books are left,
   so $278 - 121 = 157$ fiction books were borrowed.
   [3 marks available — 1 mark for working to find the total number of books left in the library, 1 mark for finding 44% of total books left in the library and 1 mark for the correct final answer]
   If you made a mistake when adding up the number of books left in the library, but did the rest of the working correctly, you'd get 2 marks out of 3 here. Questions in your exam are marked like this — so it's super important that you show all your working.

4 The multiplier for a 20% increase = 1.2 *[1 mark]*
   $1.2 \times \$927 = \$1112.40$ *[1 mark]*
   [2 marks available in total — as above]

5 a) Original price = $180, change in price = $5.40
      Percentage increase = $\frac{\$5.40}{\$180} \times 100 = 3\%$
      [2 marks available — 1 mark for using the correct formula, 1 mark for the correct answer]

   b) Buying tickets in 1 transaction:
      $(3 \times \$180) + \$5.40 = \$545.40$
      Buying tickets in 3 transactions:
      $(\$180 + \$5.40) \times 3 = \$556.20$
      Decrease in cost = $\$556.20 - \$545.40 = \$10.80$ *[1 mark]*
      Percentage decrease = $\frac{\$10.80}{\$556.20} \times 100$ *[1 mark]*
      $= 1.9417...\% = 1.94\%$ (2 d.p.) *[1 mark]*
      [3 marks available in total — as above]
      Careful here — the percentage saving is actually a percentage change, where the change is the saving and the original amount is the cost of the three separate transactions.

6 1.5% interest = $1.5 \div 100 = 0.015$
   1.5% of $326 = 0.015 \times \$326 = \$4.89$ *[1 mark]*
   Total interest = $7 \times \$4.89 = \$34.23$ *[1 mark]*
   Total amount in account = $\$326 + \$34.23 = \$360.23$ *[1 mark]*
   [3 marks available in total — as above]

7 Original volume of prism = area of cross-section × length
   $= \frac{1}{2} \times x \times x \times y = 0.5x^2y$ cm³
   Volume after increase = $\frac{1}{2} \times 1.15x \times 1.15x \times y$ *[1 mark]*
   $= 1.15^2 \times 0.5x^2y$ cm³
   $= 1.3225 \times 0.5x^2y$ cm³ *[1 mark]*
   $= 1.3225 \times$ original volume
   So the percentage increase in the volume is 32.25% *[1 mark]*
   [3 marks available in total — as above]

8 $\$28\,728 = 108\%$ *[1 mark]*
   $\$28\,728 \div 108 = \$266 = 1\%$
   $\$266 \times 100 = 100\%$ *[1 mark for dividing by 108 and multiplying by 100 (or for dividing by 1.08)]*
   $= \$26\,600$ *[1 mark]*
   [3 marks available in total — as above]

9 a) $32 is a 60% profit so $32 = 160% of cost price *[1 mark]*
      1% of cost price = $\$32 \div 160$,
      so 100% of cost price = $\$32 \div 160 \times 100$ *[1 mark]*
      = $20 *[1 mark]*
      [3 marks available — as above]

   b) Profit = $\$37.60 - \$20$ *[1 mark]* = $17.60
      % profit = $17.60 \div 20 \times 100$ *[1 mark]*
      = 88% *[1 mark]*
      [3 marks available — as above]

Pages 15-16: Compound Growth and Decay

1. a) When it was first opened $t = 0$, so the balance would have been
   $B = 5000 \times 1.02^0 = 5000 \times 1 = \$5000$ *[1 mark]*
   b) After 7 years there would be:
   $B = 5000 \times 1.02^7$ *[1 mark]*
   $= \$5743.4283...$
   $= \$5743.43$ (to the nearest cent) *[1 mark]*
   [2 marks available in total — as above]

2. Compound Collectors Account:
   Multiplier $= 1 + 0.055 = 1.055$
   $\$10\,000 \times (1.055)^5$ *[1 mark]* $= \$13\,069.60$ (2 d.p.) *[1 mark]*
   Simple Savers Account:
   6.2% of $\$10\,000 = 0.062 \times \$10\,000 = \$620$ *[1 mark]*
   $5 \times \$620 = \$3100$
   $\$10\,000 + \$3100 = \$13\,100$ so using the Simple Savers Account will give him the most money after 5 years. *[1 mark]*
   [4 marks available in total — as above]

3. $\$2704 = \$2500 \times (\text{Multiplier})^2$ *[1 mark]*
   $\dfrac{\$2704}{\$2500} = (\text{Multiplier})^2$
   Multiplier $= \sqrt{\dfrac{\$2704}{\$2500}} = 1.04$ *[1 mark]*
   Interest rate $= 1.04 - 1 = 0.04 = 4\%$ *[1 mark]*
   [3 marks available in total — as above]

4. Multiplier $= 1 - 0.29 = 0.71$
   $N_0 \times (0.71)^4 = \$482.18$ *[1 mark]*
   $N_0 = \$482.18 \div (0.71)^4$ *[1 mark]* $= \$1897.4738...$
   $= \$1897.47$ *[1 mark]*
   [3 marks available in total — as above]

5. a) Multiplier $= 1 - 0.08 = 0.92$
   Population after 15 years $= 2000 \times (0.92)^{15}$ *[1 mark]*
   $= 572.59... \approx 573$ fish *[1 mark]*
   [2 marks available in total — as above]
   b) Three quarters of initial population $= 2000 \times \dfrac{3}{4} = 1500$
   $2000 \times 0.92 = 1840$
   $2000 \times 0.92^2 = 1692.8$
   $2000 \times 0.92^3 = 1557.376$
   $2000 \times 0.92^4 = 1432.78592 < 1500$
   Population is less than $\dfrac{3}{4}$ of the initial population after 4 years.
   [2 marks available — 1 mark for calculating 2000×0.92^n for $n > 1$, 1 mark for the correct answer]

Pages 17-18: Ratios

1. a) $3\dfrac{3}{4} : 1\dfrac{1}{2} = 4 \times 3\dfrac{3}{4} : 4 \times 1\dfrac{1}{2} = 15 : 6$ *[1 mark]*
   $= 5 : 2$ *[1 mark]*
   [2 marks available in total — as above]
   b) $5 + 2 = 7$ parts
   1 part: 2800 ml $\div 7 = 400$ ml
   Yellow paint $= 400$ ml $\times 5 = 2000$ ml
   Blue paint $= 400$ ml $\times 2 = 800$ ml
   [2 marks available — 1 mark for finding the amount for 1 part, 1 mark for finding the correct amounts for both yellow and blue paint]
   If your answer to part a) was incorrect, but your answers to part b) were correct for your incorrect ratio, you still get the marks for part b).

2. a) $\dfrac{2}{9}$ as much milk is used as ice cream *[1 mark]*
   b) 1 part: $801 \div 9 = 89$ ml *[1 mark]*
   $9 + 2 = 11$ parts
   $89 \times 11 = 979$ ml *[1 mark]*
   [2 marks available in total — as above]

3. Longest $-$ shortest $= 7 - 5 = 2$ parts $= 9$ cm
   1 part $= 9 \div 2 = 4.5$ cm *[1 mark]*
   Original piece of wood is $5 + 6 + 6 + 7 = 24$ parts
   So, the original piece of wood is $24 \times 4.5 = 108$ cm *[1 mark]*
   [2 marks available in total — as above]

4. 250 ml bottle: $250 \div 200 = 1.25$ ml per cent
   330 ml bottle: $330 \div 275 = 1.2$ ml per cent
   525 ml bottle: $525 \div 375 = 1.4$ ml per cent
   So the 525 ml bottle is the best value for money.
   [3 marks available — 3 marks for finding the correct amounts per cent for all three bottles and the correct answer, otherwise 2 marks for two correct amounts per cent or 1 mark for one correct amount per cent]
   You could also compare the cost per ml of each bottle.

5. Catrin, Ariana, Nasir and Simone shared the money in the ratio $1 : 2 : 4 : 8$ *[1 mark for $1 : 2 : 4 : 8$ or an equivalent ratio (order may be different)]*
   $660 \div (1 + 2 + 4 + 8) = 44$ *[1 mark]*
   Simone got $\$44 \times 8 = \$352$ *[1 mark]*
   [3 marks available in total — as above]
   You could answer this question using a formula — if you let x be the amount of money that Catrin gets, then $x + 2x + 4x + 8x = \$660$.

6. Wolves : Bears $= (7 : 4) \times 11 = 77 : 44$
   Wolves : Giraffes $= (11 : 8) \times 7 = 77 : 56$
   Bears : Giraffes $= 44 : 56 = 11 : 14$
   [2 marks available — 1 mark for finding ratios with a common number of wolves, 1 mark for the correct answer]

Pages 19-20: Proportion

1. She worked 28 hours and got $\$231$ so she gets paid
   $\$231 \div 28 = \$8.25$ per hour *[1 mark]*.
   So in total, she'll get paid
   $\$231 + (3 \times 25 \times \$8.25) = \$849.75$ *[1 mark]*.
   [2 marks available in total — as above]

2. 1 T-shirt will take: 5 m$^2 \div 8 = 0.625$ m^2 of cotton *[1 mark]*
   85 T-shirts will take: 0.625 m$^2 \times 85 = 53.125$ m^2 of cotton *[1 mark]*
   53.125 m^2 of cotton costs $\$2.75 \times 53.125 = \$146.09375$
   $= \$146.09$ *[1 mark]*
   [3 marks available in total — as above]

3. 12 people take 3 hours.
   1 person will take $3 \times 12 = 36$ hours.
   4 people will take $36 \div 4 = 9$ hours.
   [2 marks available — 1 mark for a correct method, 1 mark for the correct answer]
   Alternatively, there are a third of the people ($12 \div 4 = 3$) so it will take three times as long — $3 \times 3 = 9$ hours.

4. To knit the same amount as Yusef, Sophie will take:
   26.5 hours $\div 2 = 13.25$ hours *[1 mark]*
   Sophie needs to knit three times as many so it will take her:
   $13.25 \times 3 = 39.75$ hours *[1 mark]*
   [2 marks available in total — as above]

5. a) For 1 sponge cake she'd need:
   Flour: 275 g $\div 5 = 55$ g   Butter: 275 g $\div 5 = 55$ g
   Sugar: 220 g $\div 5 = 44$ g   Eggs: $5 \div 5 = 1$ egg
   So for 18 sponge cakes she'll use:
   Flour: 55 g $\times 18 = 990$ g   Butter: 55 g $\times 18 = 990$ g
   Sugar: 44 g $\times 18 = 792$ g   Eggs: $1 \times 18 = 18$ eggs
   [3 marks available — 1 mark for dividing the quantities by 5, 1 mark for multiplying the quantities by 18, 1 mark for all four correct answers]
   An alternative method is to find the multiplier from 5 to 18 (it's $18 \div 5 = 3.6$) and multiply each quantity in the original recipe by this number.
   b) There will be a total of $18 \times 10 = 180$ slices *[1 mark]*
   At 50c each this will make:
   $180 \times 50c = 9000c = \$90$ *[1 mark]*
   Profit $= \$90 - \$25.30 = \$64.70$ *[1 mark]*
   [3 marks available in total — as above]

6. a) 250 people can be catered for 6 days
   1 person can be catered for $6 \times 250 = 1500$ days
   300 people can be catered for $1500 \div 300 = 5$ days
   [2 marks available — 1 mark for a correct method, 1 mark for the correct answer]

Answers

b) For a 1-day cruise it could cater for 6 × 250 = 1500 people
For a 2-day cruise it could cater for 1500 ÷ 2 = 750 people
So it can cater for 750 – 250 = 500 more people
[3 marks available — 1 mark for a correct method to find the number of people catered for on a 2-day cruise, 1 mark for the correct number of people catered for on a 2-day cruise, 1 mark for the correct final answer]

Page 21: Rounding Numbers

1 E.g. Height of man = 176 cm ≈ 180 cm
   Height of penguin ≈ 180 ÷ 3 *[1 mark]*
   = 60 cm (accept 50-67 cm) *[1 mark]*
   [2 marks available in total — as above]

2 430 light years *[1 mark]*

3 a) E.g. ($4.95 × 28) + ($11 × 19) ≈ ($5 × 30) + ($10 × 20)
      = $150 + $200 = $350
      [2 marks available — 1 mark for rounding each value sensibly, 1 mark for a sensible estimate]
   b) E.g. This is a sensible estimate as it is very close to the actual value of $347.60 *[1 mark]*.

4 $\sqrt{\frac{2321}{19.673 \times 3.81}} \approx \sqrt{\frac{2000}{20 \times 4}}$ *[1 mark]*
   $= \sqrt{\frac{100}{4}} = \sqrt{25} = 5$ *[1 mark]*
   [2 marks available in total — as above]

Page 22: Bounds

1 Minimum weight = 56.5 kg *[1 mark]*
   Maximum weight = 57.5 kg *[1 mark]*
   [2 marks available in total — as above]

2 Upper bound of x = 2.25 *[1 mark]*
   Lower bound of x = 2.15 *[1 mark]*
   Written as an interval, this is $2.15 \leq x < 2.25$
   [2 marks available in total — as above]

3 Lower bound of difference = 13.65 – 8.35 *[1 mark]*
                              = 5.3 litres *[1 mark]*
   [2 marks available in total — as above]

4 Upper bound of area = 5.25 cm² *[1 mark]*
   Lower bound of height = 3.15 cm *[1 mark]*
   2 × (5.25 ÷ 3.15) = 3.33 cm to 2 d.p. *[1 mark]*
   [3 marks available — as above]

Pages 23-24: Standard Form

1 907 200 000 = 9.072×10^8 *[1 mark]*

2 1.27×10^4 = 12 700 *[1 mark]*
   = twelve thousand seven hundred kilometres *[1 mark]*
   [2 marks available in total — as above]

3 time (s) = distance (miles) ÷ speed (miles/s)
            = $(9.3 \times 10^7) \div (1.86 \times 10^5)$ seconds
            = $(9.3 \div 1.86) \times 10^2$ seconds or 500 seconds *[1 mark]*
            = 5×10^2 seconds *[1 mark]*
   [2 marks available in total — as above]

4 a) Particle C *[1 mark]*
   b) 1.4×10^{-6} = 0.0000014 g *[1 mark]*
   c) $(3.2 \times 10^{-7}) - (2.1 \times 10^{-7}) = (3.2 - 2.1) \times 10^{-7}$
      $= 1.1 \times 10^{-7}$ g *[1 mark]*

5 a) number of tablets = dose (grams) ÷ dose per tablet (grams)
      $= (4 \times 10^{-4}) \div (8 \times 10^{-5})$ *[1 mark]*
      $= (4 \div 8) \times (10^{-4} \div 10^{-5})$
      $= 0.5 \times 10^1$ *[1 mark]*
      = 5 *[1 mark]*
      [3 marks available in total — as above]
   b) new dose = 4×10^{-4} grams + 6×10^{-5} grams *[1 mark]*
      = 4×10^{-4} grams + 0.6×10^{-4} grams *[1 mark]*
      = $(4 + 0.6) \times 10^{-4}$ grams
      = 4.6×10^{-4} grams per day *[1 mark]*
      [3 marks available in total — as above]
      You could write 4×10^{-4} as 40×10^{-5} and add it to 6×10^{-5}.

6 $(4.5 \times 10^9) \div (1.5 \times 10^8) = 30$ *[1 mark]*.
   So the ratio is 1 : 30 *[1 mark]*.
   [2 marks available in total — as above]

7 $\frac{3^2}{2^{122} \times 5^{120}} = \frac{9}{2^2(2^{120} \times 5^{120})} = \frac{9}{2^2 \times 10^{120}} = \frac{9}{4} \times \frac{1}{10^{120}} = 2.25 \times 10^{-120}$
   [2 marks available —1 mark for writing the denominator as a multiple of a power of 10, 1 mark for the correct answer]

Pages 25-27: Sets and Venn Diagrams

1 a) The universal set is the elements {1, 2, 3, 4, 5, 6, 7} *[1 mark]*.
   b) There are 15 elements within the set $(A \cup B)$ *[1 mark]*.
   c) Set J is a subset of the set $(K \cap L)$ *[1 mark]*.
   d) Set Z is an empty set, as there are no square numbers between 5 and 8 *[1 mark]*.

2 Elements of A = {3, 4 , 6, 9, 12}
   Elements of B = {4, 9, 16}
   Elements of $A \cap B$ = {4, 9} *[1 mark]*

3 Number of elements in Z and in Y = 45 – 21 = 24
   Number of elements in Y but not in Z = 39 – 24 = 15
   Number of elements not in Y or Z = 60 – 24 – 21 – 15 = 0

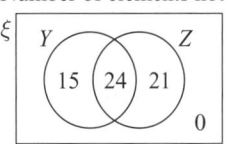

   [2 marks available — 2 marks for a fully correct Venn diagram, otherwise 1 mark for two correct values]

4 $(x + 4) + 3x + (2x + 3) + 9 = 40$
   $6x + 16 = 40$ *[1 mark]*
   $6x = 24$
   $x = 4$ *[1 mark]*
   [2 marks available in total — as above]

5 Number of elements not in R or W = 40
   Number of elements in either R or W = 400 – 40 = 360
   Number of elements in R and W = (300 + 310) – 360 = 250
   Number of elements in R but not in W = 300 – 250 = 50
   Number of elements in W but not in R = = 310 – 250 = 60

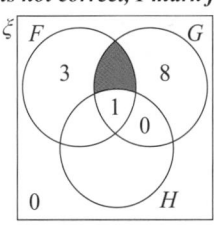

   [2 marks available — 2 marks for a fully correct Venn diagram, otherwise 1 mark for two correct values]

6 a) E = {5, 6, 9, 10, 15, 18, 30, 45}
      $(C' \cap D \cap E)$ = {6, 9, 18}
      [2 marks available — 2 marks for all three correct numbers only, otherwise 1 mark for any two correct numbers]
   b) $(C \cup D \cup E)'$ is the complement of $(C \cup D \cup E)$.
      6, 9 and 18 would be elements of $(C \cup D \cup E)$,
      so they would not be elements of $(C \cup D \cup E)'$.
      This means that Shannon is not correct.
      [2 marks available — 1 mark for stating that Shannon is not correct, 1 mark for at least one piece of evidence]

7 a)

   [1 mark]
   b) $n(F \cap G \cap H')$ = 15 – 8 – 1 – 0 = 6 *[1 mark]*
      $n(F \cap G' \cap H)$ = 14 – 3 – 6 – 1 = 4 *[1 mark]*
      [2 marks available — as above]

8   Elements of P = {2, 3, 4, 5, 6, 7, 8}
    Elements of P' = {1, 9, 10}
    Elements of Q = {3, 6, 9}
    Elements of R = {2, 4, 6, 8, 10}
    Elements of R' = {1, 3, 5, 7, 9}
    $(P \cap Q)$ = {3, 6}
    3 is an element of $(P \cap Q)$ but not an element of R,
    so $(P \cap Q) \not\subseteq R$ — Barry is incorrect.
    $(P' \cap Q)$ = {9}
    All of the elements of $(P' \cap Q)$ are elements of R'.
    So $(P' \cap Q) \subseteq R'$ — Michiko is correct.
    [4 marks available — 1 mark for the correct elements of $(P \cap Q)$, 1 mark for the correct elements of $(P' \cap Q)$, 1 mark for finding Barry is incorrect, 1 mark for finding that Michiko is correct]

Section Two — Algebra

Page 28: Algebra Basics

1   10s *[1 mark]*
2   a) 4p *[1 mark]*
    b) 4p + 3r
        [2 marks available — 1 mark for 4p and 1 mark for 3r]
    c) $x^2 + 4x$
        [2 marks available — 1 mark for x^2 and 1 mark for 4x]
3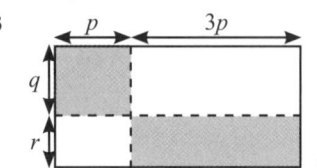
    [1 mark]
4   Perimeter of rectangle = 4x + 3 + 4x + 3 + 5x – 9 + 5x – 9
    = 18x – 12 cm *[1 mark]*.
    So perimeter of hexagon = 18x – 12 cm.
    Hexagon side length = (18x – 12) ÷ 6 *[1 mark]*
    = 3x – 2 cm *[1 mark]*
    [3 marks available in total — as above]

Page 29: Powers

1   $5^{-2} = \frac{1}{5^2} = \frac{1}{25}$ *[1 mark]*
2   $y^{-3} = \frac{1}{y^3}$, $y^1 = y$, $y^0 = 1$
    so the correct order is: y^{-3} y^0 y^1 y^3
    [2 marks available — 2 marks for all 4 in the correct order, otherwise 1 mark for any 3 in the correct relative order]
    If you can't identify which term is the smallest just by looking at them, try substituting a value for y into all the expressions and working out the answer. Then it'll be easy to tell which is the smallest.
3   $\frac{3^4 \times 3^7}{(3^6)^{-2}} = \frac{3^{(4+7)}}{(3^6)^{-2}} = \frac{3^{11}}{3^{-12}} = 3^{(11--12)} = 3^{23}$
    [2 marks available — 1 mark for a correct attempt at adding or subtracting powers, 1 mark for the correct final answer]
4   $(x^4 \times x^7) = x^{(4+7)} = x^{11}$
    $(x^3 \times x^2) = x^{(3+2)} = x^5$, so $(x^3 \times x^2)^2 = (x^5)^2 = x^{10}$
    So $(x^4 \times x^7) \div (x^3 \times x^2)^2 = x^{11} \div x^{10} = x^1 = x$
    [2 marks available — 1 mark if each bracket has been correctly simplified, 1 mark for the correct answer]
5   $16^{\frac{1}{4}} = \sqrt[4]{16} = 2$ *[1 mark]*
    So $16^{\frac{3}{4}} = 2^3 = 8$ *[1 mark]*
    [2 marks available in total — as above]
6   a) $\frac{2}{5}x^{\frac{1}{2}} \div 2x^{-2} = \left(\frac{2}{5} \div 2\right)x^{\frac{1}{2}--2} = \frac{1}{5}x^{\frac{5}{2}}$
        [2 marks available — 1 mark for correct attempt at subtracting powers, 1 mark for the correct final answer]
    b) $\left(\frac{24a^{30}}{3a^3}\right)^{\frac{1}{3}} = (8a^{(30-3)})^{\frac{1}{3}} = (8a^{27})^{\frac{1}{3}} = 2a^9$
        [3 marks available — 1 mark for simplifying the fraction to $8a^{27}$, 1 mark for a coefficient of 2 in the final answer, 1 mark for a^9 in the final answer]

Page 30: Expanding Brackets

1   a) $5p(6 – 2p) = 30p – 10p^2$
        [2 marks available — 1 mark for each term]
    b) $(2t – 5)(3t + 4) = (2t \times 3t) + (2t \times 4) + (–5 \times 3t) + (–5 \times 4)$
        $= 6t^2 + 8t – 15t – 20$ *[1 mark]* $= 6t^2 – 7t – 20$ *[1 mark]*
        [2 marks available in total — as above]
2   $4(5x – 7) + 6(4 – 2x) = 20x – 28 + 24 – 12x = 8x – 4$
    So a, b and c = 1, 8 and –4 (or 2, 4 and –2) (or 4, 2 and –1)
    [3 marks available — 1 mark for each correct value]
3   Area of rectangle = $(4x + 4y) \times (3x – 2y)$
    $= 12x^2 – 8xy + 12xy – 8y^2$ *[1 mark]*
    $= 12x^2 + 4xy – 8y^2$ *[1 mark]*
    [2 marks available in total — as above]
4   $(5c + 6)(2c + 1)(c + 1) = (5c + 6)(2c^2 + 3c + 1)$
    $= 5c \times (2c^2 + 3c + 1) + 6 \times (2c^2 + 3c + 1)$
    $= (10c^3 + 15c^2 + 5c) + (12c^2 + 18c + 6)$
    $= 10c^3 + 27c^2 + 23c + 6$
    [3 marks available — 1 mark for expanding two brackets correctly, 1 mark for correctly multiplying by the remaining bracket, 1 mark for simplifying the cubic equation]
    It's fine to multiply any pair of brackets in the first step. If you follow the method, you should always end up with the same answer.

Page 31: Factorising

1   a) $7y – 21y^2 = 7(y – 3y^2) = 7y(1 – 3y)$
        [2 marks available — 1 mark for 7y, 1 mark for (1 – 3y)]
    b) $8ab^2 – 12a^2bc = 4(2ab^2 – 3a^2bc) = 4ab(2b – 3ac)$
        [2 marks available — 1 mark for 4ab, 1 mark for (2b – 3ac)]
    c) $12pr + 6qr – 30pqr = 6r(2p + q – 5pq)$
        [2 marks available — 1 mark for 6r, 1 mark for (2p + q – 5pq)]
2   a) $x^2 + 6xy + 9y^2 = (x + 3y)(x + 3y) = (x + 3y)^2$ *[1 mark]*
    b) $16p^2 – 9q^2 = (4p – 3q)(4p + 3q)$
        [2 marks available — 2 marks for the correct final answer, otherwise 1 mark for attempting to use the difference of two squares]
    c) $3g – 6h + 4fg – 8fh = g(3 + 4f) – 2h(3 + 4f)$
        $= (g – 2h)(3 + 4f)$
        [2 marks available — 1 mark for factorising g and h separately, 1 mark for the correct final answer]

Page 32: Manipulating Surds

1   $(2 + \sqrt{3})(5 – \sqrt{3}) = (2 \times 5) + (2 \times –\sqrt{3}) + (\sqrt{3} \times 5) + (\sqrt{3} \times –\sqrt{3})$
    $= 10 – 2\sqrt{3} + 5\sqrt{3} – 3 = 7 + 3\sqrt{3}$
    [2 marks available — 1 mark for correct working, 1 mark for the correct answer]
2   $2\sqrt{50} = 2\sqrt{25 \times 2} = 2 \times 5\sqrt{2} = 10\sqrt{2}$
    $(\sqrt{2})^3 = \sqrt{2} \times \sqrt{2} \times \sqrt{2} = (\sqrt{2})^2 \times \sqrt{2} = 2\sqrt{2}$
    So $2\sqrt{50} – (\sqrt{2})^3 = 10\sqrt{2} – 2\sqrt{2} = 8\sqrt{2}$
    [2 marks available — 2 marks for the correct answer, otherwise 1 mark for correctly simplifying either surd]
3   $\sqrt{396} = \sqrt{36 \times 11} = 6\sqrt{11}$ *[1 mark]*
    $\frac{22}{\sqrt{11}} = \frac{22\sqrt{11}}{11} = 2\sqrt{11}$ *[1 mark]*
    $\sqrt{44} = \sqrt{4 \times 11} = 2\sqrt{11}$
    So $\frac{220}{\sqrt{44}} = \frac{220}{2\sqrt{11}} = \frac{220\sqrt{11}}{22} = 10\sqrt{11}$ *[1 mark]*
    So $\sqrt{396} + \frac{22}{\sqrt{11}} – \frac{220}{\sqrt{44}} = 6\sqrt{11} + 2\sqrt{11} – 10\sqrt{11} = –2\sqrt{11}$
    [1 mark]
    [4 marks available — as above]
4   $\frac{1 + \sqrt{7}}{3 – \sqrt{7}} = \frac{(1 + \sqrt{7})(3 + \sqrt{7})}{(3 – \sqrt{7})(3 + \sqrt{7})}$ *[1 mark]*
    $= \frac{3 + \sqrt{7} + 3\sqrt{7} + 7}{9 – 7}$ *[1 mark]*
    $= \frac{10 + 4\sqrt{7}}{2}$ *[1 mark]* $= 5 + 2\sqrt{7}$ *[1 mark]*
    [4 marks available in total — as above]

Pages 33-34: Solving Equations

1. a) $p - 11 = -7$, so $p = 4$ *[1 mark]*
   b) $3z + 2 = z + 15$
      $2z = 13$ *[1 mark]*
      $z = 13 \div 2 = 6.5$ *[1 mark]*
      [2 marks available in total — as above]

2. a) $3(a + 2) = 15$
      $3a + 6 = 15$
      $3a = 9$ *[1 mark]*
      $a = 3$ *[1 mark]*
      [2 marks available in total — as above]
   b) $5(2b - 1) = 4(3b - 2)$
      $10b - 5 = 12b - 8$ *[1 mark]*
      $3 = 2b$ *[1 mark]*
      $b = 1.5$ *[1 mark]*
      [3 marks available in total — as above]

3. $(x + 2)(x - 4) = (x - 2)(x + 1)$
   $x^2 - 2x - 8 = x^2 - x - 2$
   $-8 + 2 = -x + 2x$
   $-6 = x$ so $x = -6$
   [4 marks available — 1 mark for expanding the brackets on the RHS, 1 mark for expanding the brackets on the LHS, 1 mark for collecting like terms on each side, 1 mark for the correct solution]

4. a) $5x^2 = 180$
      $x^2 = 36$ *[1 mark]*
      $x = \pm 6$ *[1 mark]*
      [2 marks available in total — as above]
   b) $\dfrac{8 - 2x}{3} + \dfrac{2x + 4}{9} = 12$
      $\dfrac{9(8 - 2x)}{3} + \dfrac{9(2x + 4)}{9} = 9 \times 12$
      $3(8 - 2x) + (2x + 4) = 108$
      $24 - 6x + 2x + 4 = 108$
      $6x - 2x = 24 + 4 - 108$
      $4x = -80$ so $x = -20$
      [4 marks available — 2 marks for rearranging to remove the fractions, 1 mark for rearranging to get all x-terms on one side, 1 mark for correct answer]

5. $2x + 6 = 5(x - 3)$ *[1 mark]*
   $2x + 6 = 5x - 15$
   $21 = 3x$ *[1 mark]*
   $x = 7$ *[1 mark]*
   So one side of the triangle measures $2(7) + 6 = 20$ cm *[1 mark]*
   [4 marks available in total — as above]

6. If Neil worked h hours, Liam worked $(h + 30)$ hours.
   $360 \div 4.5 = 80$
   $80 = h + (h + 30) = 2h + 30$
   $50 = 2h$, so $h = 25$
   Neil worked 25 hours and Liam worked $(25 + 30) = 55$ hours.
   [3 marks available — 1 mark for forming an equation for the total number of hours, 1 mark for solving the equation, 1 mark for finding the number of hours each boy worked]

Pages 35-36: Formulas

1. a) $u = v - at$ *[1 mark]*
   b) $v - u = at$ *[1 mark]*
      $t = \dfrac{v - u}{a}$ *[1 mark]*
      [2 marks available in total — as above]

2. a) Number of miles = (number of kilometres ÷ 8) × 5
      $m = (k \div 8) \times 5 = \dfrac{5k}{8}$
      [2 marks available — 2 marks for correct formula, otherwise 1 mark for just $\dfrac{5k}{8}$]
   b) Substitute $k = 110$ into formula: $m = \dfrac{5 \times 110}{8}$
      $m = 550 \div 8 = 68.75$
      Therefore 110 km = 68.75 miles.
      [2 marks available — 1 mark for substitution of k = 110 into formula, 1 mark for correct final answer]

3. $s = \dfrac{1}{2}gt^2$, so $gt^2 = 2s$ *[1 mark]*, $t^2 = \dfrac{2s}{g}$ *[1 mark]*,
   $t = \sqrt{\dfrac{2s}{g}}$ *[1 mark]*
   [3 marks available in total — as above]

4. Call the number of cakes Nancy bakes n. Then Chetna bakes $2n$ cakes and Norman bakes $2n + 12$ cakes. They bake 72 cakes, so
   $n + 2n + 2n + 12 = 72$
   $5n + 12 = 72$
   $5n = 60$ $n = 12$
   So Nancy bakes 12 cakes, Chetna bakes $12 \times 2 = 24$ cakes and Norman bakes $24 + 12 = 36$ cakes.
   [4 marks available — 1 mark for forming expressions for the number of cakes each person bakes, 1 mark for forming an equation for the total number of cakes baked, 1 mark for solving the equation, 1 mark for the correct numbers of cakes each person bakes]

5. a) $a + y = \dfrac{b - y}{a}$, so:
      $a(a + y) = b - y$ *[1 mark]*, $a^2 + ay = b - y$,
      $ay + y = b - a^2$ *[1 mark]*, $y(a + 1) = b - a^2$ *[1 mark]*,
      $y = \dfrac{b - a^2}{a + 1}$ *[1 mark]*
      [4 marks available in total — as above]
   b) When $a = 3$ and $b = 6$, $y = \dfrac{6 - 3^2}{3 + 1} = -\dfrac{3}{4}$ or -0.75 *[1 mark]*

6. a) $x = \sqrt{\dfrac{n - 1}{2n - 3}}$, so $x^2 = \dfrac{n - 1}{2n - 3}$ *[1 mark]*
      $x^2(2n - 3) = n - 1$, so $2nx^2 - 3x^2 = n - 1$ *[1 mark]*
      $2nx^2 - 3x^2 - n = -1$, so $2nx^2 - n = 3x^2 - 1$ *[1 mark]*
      $n(2x^2 - 1) = 3x^2 - 1$ *[1 mark]*, so $n = \dfrac{3x^2 - 1}{2x^2 - 1}$ *[1 mark]*
      [5 marks available in total — as above]
   b) $n = \dfrac{(3(0.2)^2 - 1)}{(2(0.2)^2 - 1)} = 0.956521... = 0.957$ (3 s.f.) *[1 mark]*

Page 37: Algebraic Fractions

1. a) $\dfrac{x^2 - 2x}{x^2 - 5x + 6} = \dfrac{x(x - 2)}{(x - 2)(x - 3)} = \dfrac{x}{(x - 3)}$
      [3 marks available — 1 mark for correctly factorising the denominator, 1 mark for correctly factorising the numerator, 1 mark for the correct answer]
   b) $\dfrac{4x^2 + 10x - 6}{16x^2 - 4} = \dfrac{2(2x^2 + 5x - 3)}{4(4x^2 - 1)}$
      $= \dfrac{2(2x - 1)(x + 3)}{4(2x + 1)(2x - 1)} = \dfrac{x + 3}{2(2x + 1)}$
      [3 marks available — 1 mark for correctly factorising the denominator, 1 mark for correctly factorising the numerator, 1 mark for the correct answer]

2. $\dfrac{x + 4}{2x + 1} + \dfrac{2x - 2}{x - 2} = \dfrac{(x + 4)(x - 2)}{(2x + 1)(x - 2)} + \dfrac{(2x - 2)(2x + 1)}{(2x + 1)(x - 2)}$
   $\dfrac{x^2 - 2x + 4x - 8}{(2x + 1)(x - 2)} + \dfrac{4x^2 + 2x - 4x - 2}{(2x + 1)(x - 2)} = \dfrac{5x^2 - 10}{(2x + 1)(x - 2)}$
   [4 marks available — 1 mark for using the correct common denominator, 1 mark for correctly expanding the first numerator, 1 mark for correctly expanding the second numerator, 1 mark for the correct answer]

3. $\dfrac{2a - 8}{a^2 - 9} \div \dfrac{a^2 - 2a - 8}{a^2 + 5a + 6} \times (2a^2 - a - 15)$
   $= \dfrac{2a - 8}{a^2 - 9} \times \dfrac{a^2 + 5a + 6}{a^2 - 2a - 8} \times (2a^2 - a - 15)$
   $= \dfrac{2(a - 4)}{(a + 3)(a - 3)} \times \dfrac{(a + 3)(a + 2)}{(a + 2)(a - 4)} \times (2a + 5)(a - 3)$
   $= 2(2a + 5)$ (or $4a + 10$)
   [5 marks available — 1 mark for converting to a multiplication, 1 mark for factorising the first fraction, 1 mark for factorising the second fraction, 1 mark for factorising the quadratic term, 1 mark for cancelling to reach correct answer]

Pages 38-39: Factorising Quadratics

1. 3 and 6 multiply to give 18 and add to give 9,
   so $x^2 + 9x + 18 = (x + 3)(x + 6)$
   [2 marks available — 1 mark for correct numbers in brackets, 1 mark for correct signs]
   The brackets can be either way around —
   $(x + 6)(x + 3)$ is also correct.

2. 1 and 5 multiply to give 5 and subtract to give –4,
   so $y^2 - 4y - 5 = (y + 1)(y - 5)$
   [2 marks available — 1 mark for correct numbers in brackets, 1 mark for correct signs]

3. Take out a multiple of x to give $x(x^2 + 4x - 32)$,
   then factorise the bracket as normal:
   4 and 8 multiply to give 32 and subtract to give 4,
   so $x(x^2 + 4x - 32) = x(x - 4)(x + 8)$
   [3 marks available — 1 mark for correct numbers in brackets, 1 mark for correct signs, 1 mark for correct factorised expression]

4. 6 and 2 multiply to give 12 and subtract to give 4,
   so if $x^2 + 4x - 12 = 0$,
   $(x + 6)(x - 2) = 0$
   $x + 6 = 0$ or $x - 2 = 0$
   $x = -6$ or $x = 2$
   [3 marks available — 1 mark for correct numbers in brackets, 1 mark for correct signs, 1 mark for both solutions]

5. 1 and 7 multiply to give 7, and (2×7) and 1 subtract to give 13,
   so $2x^2 + 13x - 7 = (2x - 1)(x + 7)$
   $2x - 1 = 0$ or $x + 7 = 0$
   $x = 0.5$ or $x = -7$
   [3 marks available — 1 mark for correct numbers in brackets, 1 mark for correct signs, 1 mark for both solutions correct]

6. $3x = \frac{15 - 9x}{2x}$
   $6x^2 + 9x - 15 = 0$
   $2x^2 + 3x - 5 = 0$
   1 and 5 multiply to give 5, and 5 and (2×1) subtract to give 3,
   so $2x^2 + 3x - 5 = (2x + 5)(x - 1)$
   $2x + 5 = 0$ or $x - 1 = 0$
   $x = -\frac{5}{2}$ or $x = 1$
   [4 marks available — 1 mark for forming the correct quadratic equation, 1 mark for correct numbers in brackets, 1 mark for correct signs, 1 mark for both solutions]

7. Let the two consecutive even numbers be $2n$ and $2n + 2$.
   $2n(2n + 2) = 288$ *[1 mark]*
   $4n^2 + 4n = 288$
   $4n^2 + 4n - 288 = 0$
   $n^2 + n - 72 = 0$
   $(n + 9)(n - 8) = 0$ *[1 mark]*, so $n = -9$ or $n = 8$.
   The numbers are positive, so $n = 8$ *[1 mark]*.
   The larger of the two numbers is $2(8) + 2 = 18$ *[1 mark]*.
   [4 marks available in total — as above]
   You could also solve $n(n + 2) = 288$ to get $n = 16$ and the larger number as $n + 2 = 18$.

Page 40: The Quadratic Formula

1. $a = 2$, $b = -7$ and $c = 2$
   $x = \frac{-(-7) \pm \sqrt{(-7)^2 - 4 \times 2 \times 2}}{2 \times 2} = \frac{7 \pm \sqrt{33}}{4}$
   $x = 3.186140...$ or $x = 0.3138593... = 3.19$ or 0.31 (2 d.p.)
   [3 marks available — 1 mark for correct substitution, 1 mark for each correct solution]

2. $a = 3$, $b = -2$ and $c = -4$
   $x = \frac{-(-2) \pm \sqrt{(-2)^2 - 4 \times 3 \times -4}}{2 \times 3} = \frac{2 \pm \sqrt{52}}{6}$
   $x = 1.535183...$ or $x = -0.8685170... = 1.54$ or -0.87 (2 d.p.)
   [3 marks available — 1 mark for correct substitution, 1 mark for each correct solution]

3. $6x^2 + 4x - 3 = 8$, so $6x^2 + 4x - 11 = 0$
   $a = 6$, $b = 4$ and $c = -11$
   $x = \frac{-4 \pm \sqrt{4^2 - 4 \times 6 \times -11}}{2 \times 6} = \frac{-4 \pm \sqrt{280}}{12}$
   $x = 1.061100...$ or $x = -1.727766... = 1.06$ or -1.73 (2 d.p.)
   [3 marks available — 1 mark for correct substitution, 1 mark for each correct solution]

4. $(x + 3)(3x + 3) = 30$ *[1 mark]*
   $3x^2 + 12x + 9 = 30$
   $3x^2 + 12x - 21 = 0$ *[1 mark]*
   $x^2 + 4x - 7 = 0$
   $a = 1$, $b = 4$ and $c = -7$
   $x = \frac{-4 \pm \sqrt{4^2 - 4 \times 1 \times -7}}{2 \times 1}$ *[1 mark]* $= \frac{-4 \pm \sqrt{44}}{2}$
   $= 1.31662...$ or $-5.31662...$
   Lengths cannot be negative, so $x = 1.31662...$ *[1 mark]*
   So the longer side is $3(1.31662...) + 3 = 6.9498...$
   $= 6.95$ cm (2 d.p.) *[1 mark]*
   [5 marks available in total — as above]

Page 41: Completing the Square

1. $(x + 2)^2 - 9 = x^2 + 4x + 4 - 9$ *[1 mark]* $= x^2 + 4x - 5$
   $a = 4$ and $b = -5$ *[1 mark]*
   [2 marks available in total — as above]

2. a) $-10 \div 2 = -5$, so $p = -5$ and the bit in brackets is $(x - 5)^2$. *[1 mark]*
   Expanding the brackets: $(x - 5)^2 = x^2 - 10x + 25$.
   To complete the square: $-5 - 25 = -30$, so $q = -30$. *[1 mark]*
   $p = -5$ and $q = -30$
   [2 marks available in total — as above]

   b) $(x - 5)^2 - 30 = 0$, so $(x - 5)^2 = 30$ and $x - 5 = \pm\sqrt{30}$
   So $x = 5 + \sqrt{30}$ or $x = 5 - \sqrt{30}$
   [2 marks available — 1 mark for each correct solution]

3. a) Take a factor of 2 out to give $2(x^2 + 3x)$, $3 \div 2 = \frac{3}{2}$,
   so $a = 2$, $s = \frac{3}{2}$ and the initial bracket $= 2\left(x + \frac{3}{2}\right)^2$ *[1 mark]*
   Expanding the brackets: $2\left(x + \frac{3}{2}\right)^2 = 2x^2 + 6x + \frac{9}{2}$.
   To complete the square: $7 - \frac{9}{2} = \frac{5}{2}$, so $t = \frac{5}{2}$.
   So the expression is $2\left(x + \frac{3}{2}\right)^2 + \frac{5}{2}$ *[1 mark]*
   [2 marks available in total — as above]

   b) The minimum point of the curve occurs when the bracket is equal to zero, so $x = -\frac{3}{2}$. *[1 mark]*
   The y value is equal to the adjusting number, so $y = \frac{5}{2}$. *[1 mark]*
   So the minimum point is at $\left(-\frac{3}{2}, \frac{5}{2}\right)$.
   [2 marks available in total — as above]

Pages 42-43: Simultaneous Equations

1. $3x + 4y = 26$ (1)
   $2x + 2y = 14$ (2) $\xrightarrow{\times 2}$ $4x + 4y = 28$ (3) *[1 mark]*
   (3) – (1):
   $4x + 4y = 28$ $3x + 4y = 26$
   $-3x + 4y = 26$ $(3 \times 2) + 4y = 26$
   $x = 2$ *[1 mark]* $4y = 26 - 6 = 20$
                               $y = 5$ *[1 mark]*
   [3 marks available in total — as above]

2. $x + 3y = 11$ (1) $\xrightarrow{\times 3}$ $3x + 9y = 33$ (3) *[1 mark]*
   $3x + y = 9$ (2)
   (3) – (2):
   $3x + 9y = 33$ $x + 3y = 11$
   $-3x + y = 9$ $x + (3 \times 3) = 11$
   $8y = 24$ $x = 11 - 9$
   $y = 3$ *[1 mark]* $x = 2$ *[1 mark]*
   [3 marks available in total — as above]

3   Let f be the number of chocolate frogs and m be
    the number of sugar mice.
    $4f + 3m = \$3.69$ and $6f + 2m = \$3.96$
    $4f + 3m = \$3.69$ $\xrightarrow{\times 2}$ $8f + 6m = \$7.38$
    $6f + 2m = \$3.96$ $\xrightarrow{\times 3}$ $18f + 6m = \$11.88$

    $18f + 6m = \$11.88$ $4f + 3m = \$3.69$
    $8f + 6m = \$7.38$ – $3m = \$3.69 – (4 \times 0.45)$
    $10f = \$4.50$ $3m = \$1.89$
    $f = \$0.45$ $m = \$0.63$
    So a bag with 2 chocolate frogs and 5 sugar mice would cost
    $(2 \times 0.45) + (5 \times 0.63) = \$4.05$
    [6 marks available in total — 1 mark for forming each equation, 1 mark for forming an equation for one of f or m, 1 mark for the correct value of f, 1 mark for the correct value of m, 1 mark for the correct final price of the bag]

4   $x^2 + y = 4$, so $y = 4 – x^2$
    $4x – 1 = 4 – x^2$ *[1 mark]*
    $x^2 + 4x – 5 = 0$ *[1 mark]*
    $(x + 5)(x – 1) = 0$ *[1 mark]*
    $x = –5$ or $x = 1$ *[1 mark]*
    When $x = 1$, $y = (4 \times 1) – 1 = 3$
    When $x = –5$, $y = (4 \times –5) – 1 = –21$
    So the solutions are $x = 1$, $y = 3$ and $x = –5$, $y = –21$ *[1 mark]*
    [5 marks available in total — as above]

5   $2x^2 – 4y = –14 – 12x$, so $4y = 2x^2 + 12x + 14$,
    so $2y = x^2 + 6x + 7$
    $x + x^2 + 6x + 7 = –5$ *[1 mark]*
    $x^2 + 7x + 12 = 0$ *[1 mark]*
    $(x + 3)(x + 4) = 0$ *[1 mark]*
    $x = –3$ or $x = –4$ *[1 mark]*
    When $x = –3$, $y = \dfrac{–5 + 3}{2} = –1$
    When $x = –4$, $y = \dfrac{–5 + 4}{2} = –\dfrac{1}{2}$
    So the solutions are $x = –3$, $y = –1$ and $x = –4$, $y = –\dfrac{1}{2}$ *[1 mark]*
    [5 marks available in total — as above]

6   $y = x^2 + 3x – 1$ and $y = 2x + 5$ so $x^2 + 3x – 1 = 2x + 5$ *[1 mark]*
    $x^2 + x – 6 = 0$
    $(x + 3)(x – 2) = 0$ *[1 mark]*
    $x = –3$ or $x = 2$ *[1 mark]*
    When $x = –3$, $y = (2 \times –3) + 5 = –6 + 5 = –1$
    When $x = 2$, $y = (2 \times 2) + 5 = 9$
    So the lines intersect at $(–3, –1)$ and $(2, 9)$ *[1 mark]*
    Change in $x = 2 – (–3) = 5$
    Change in $y = 9 – (–1) = 10$ *[1 mark for both]*
    $\sqrt{10^2 + 5^2}$ *[1 mark]* $= 11.18033... = 11.18$ (2 d.p.) *[1 mark]*
    [7 marks available in total — as above]

Pages 44-45: Solving Equations Using Graphs

1   $x = 3$ and $y = 4$ *[1 mark]*
    These are the x and y coordinates of the point where the two lines cross.

2   The graphs cross at $(–1.5, –4)$ and $(2, 3)$, so the solutions are
    $x = –1.5$, $y = –4$ *[1 mark]*
    and $x = 2$, $y = 3$ *[1 mark]*
    [2 marks available in total — as above]

3   a) $x = 1$ *[1 mark]*
    b)
    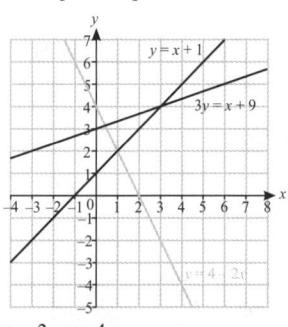
    $x = 3$, $y = 4$
    [3 marks available — 2 marks for correctly drawing the line $3y = x + 9$, 1 mark for the correct answer]

4   Find the equation of the line that should be drawn:
    $x^2 + x = 6$
    $x^2 + x – 5 = 1$
    $x^2 + 2x – 5 = x + 1$
    So draw the line $y = x + 1$ to find the solutions *[1 mark]*
    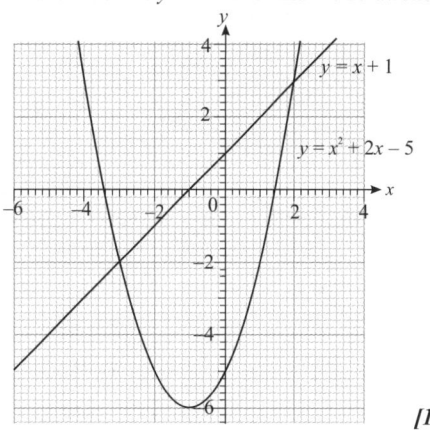
    [1 mark]
    The solutions to $x^2 + x = 6$ are:
    $x = –3$ *[1 mark]*
    $x = 2$ *[1 mark]*
    [4 marks available in total — as above]

Page 46: Inequalities

1   $–2 \leq x < 4$ *[1 mark]*

2   Largest possible value of $p = 45$
    Smallest possible value of $q = 26$ *[1 mark for both]*
    Largest possible value of $p – q = 45 – 26 = 19$ *[1 mark]*.
    [2 marks available in total — as above]

3   a) $4q – 5 < 23$
        $4q < 23 + 5$
        $4q < 28$ *[1 mark]*
        $q < 28 \div 4$
        $q < 7$ *[1 mark]*
        [2 marks available in total — as above]
    b) $4r – 2 \geq 6r + 5$
        $4r – 6r \geq 5 + 2$
        $–2r \geq 7$ *[1 mark]*
        $r \leq 7 \div –2$
        $r \leq –3.5$ *[1 mark]*
        [2 marks available in total — as above]

4   $5n – 3 \leq 17$, so $5n \leq 20$, so $n \leq 4$ *[1 mark]*
    $2n + 6 > 8$, so $2n > 2$, so $n > 1$ *[1 mark]*
    Putting these together gives $1 < n \leq 4$, so $n = \{2, 3, 4\}$ *[1 mark]*
    [3 marks available in total — as above]
    Don't forget to give your answer in set notation here.

5   $2n + (2n + 2) + (2n + 4) < 1000$ *[1 mark]*
    $6n + 6 < 1000$
    $n < 165.666...$ *[1 mark]*
    So the largest possible values of the numbers are obtained
    when $n = 165$, which gives 330, 332 and 334 *[1 mark]*.
    [3 marks available in total — as above]

Page 47: Graphical Inequalities

1   a)
    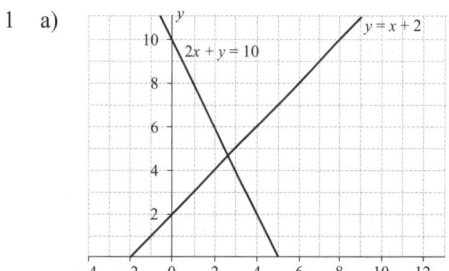
    [2 marks available — 1 mark for correctly drawing $2x + y = 10$, 1 mark for correctly drawing $y = x + 2$]

b)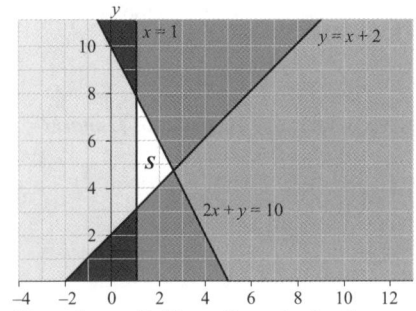

[2 marks available — 2 marks for the correct area labelled, otherwise 1 mark if the labelled area is on the wrong side of one line]

2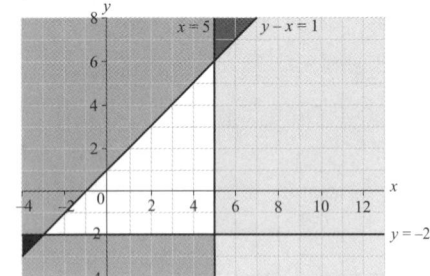

[4 marks available — 1 mark for drawing each line correctly, 1 mark for indicating the correct area]

3   $y \geq 2$ *[1 mark]*, $x + y \leq 8$ *[1 mark]* and $y \leq x$ *[1 mark]*
[3 marks available in total — as above]

Pages 48-50: Sequences

1   a)   36, 44 *[1 mark]*
  b)   23rd term = 25th term – (2 × difference between terms)
              = 196 – (2 × 8) = 196 – 16 = 180 *[1 mark]*
  c)   All terms in the sequence must be a multiple of 4 *[1 mark]*
      (the first term is 4, and the difference between the terms is 8). 90 isn't a multiple of 4, so it can't be the 12th term. *[1 mark]*
      [2 marks available in total — as above]
      You could also work out the 12th term and show that it's not 90.

2   Second term = 7 – 3 = 4
    Fourth term = 4 + 7 = 11
    Fifth term = 7 + 11 = 18
    [2 marks available — 2 marks for all three terms correct, otherwise 1 mark for at least one term correct]

3   1st term = $(1)^2 + (3 \times 1) - 2 = 2$
    2nd term = $(2)^2 + (3 \times 2) - 2 = 8$
    3rd term = $(3)^2 + (3 \times 3) - 2 = 16$
    [2 marks available — 2 marks for all three terms correct, otherwise 1 mark for two terms correct]

4   Two consecutive terms are nth and $(n + 1)$th, which have values: $3n - 10$ and $3(n + 1) - 10 = 3n - 7$.
    Their sum is $3n - 10 + 3n - 7 = 6n - 17$.
    So $6n - 17 = 223$
    $6n = 240$ and $n = 40$
    So the two terms are $(3 \times 40) - 10 = 110$ and $(3 \times 40) - 7 = 113$.
    [4 marks available — 1 mark for finding expressions for both terms, 1 mark for setting their sum equal to 223, 1 mark for solving the equation, 1 mark for both correct terms]

5   a)   Common difference = 4 so $4n$ is in the formula.
      To get from $4n$ to each term, you have to subtract 1, so the expression for the nth term is $4n - 1$.
      [2 marks available — 2 marks for correct expression, otherwise 1 mark for finding $4n$]
  b)   If 89 is a term in this sequence, then $4n - 1 = 89$ *[1 mark]*.
      Then $4n = 90$ and $n = 22.5$. n is not a whole number, so 89 is not in the sequence. *[1 mark]*
      [2 marks available in total — as above]

6   a)   The difference increases by 20 each time, so the next two terms are 360 and 490.
      [2 marks available — 1 mark for each correct term]

  b)   The terms can be compared to n^2. The terms in the sequence are 10 times the square numbers, so the expression for the nth term is $10n^2$.
      [2 marks available — 2 marks for the correct answer, otherwise 1 mark for an expression involving n^2]

7   a)   Sequence: 2   10   30   68   130
      First difference: 8   20   38   62
      Second difference: 12   18   24
      Third difference: 6   6
      Third difference = 6, so $\frac{6}{6}n^3 = n^3$ is in the formula.
      To get from n^3 to each term, you have to add n, so the expression for the nth term is $n^3 + n$.
      [2 marks available — 2 marks for correct expression, otherwise 1 mark for finding n^3]
  b)   The nth term is $n^3 + n$ and the $(n + 1)$th term is
      $(n + 1)^3 + (n + 1) = n^3 + 3n^2 + 4n + 2$
      Product of the nth and $(n + 1)$th terms:
      $(n^3 + n)(n^3 + 3n^2 + 4n + 2)$
      $= n^3 \times (n^3 + 3n^2 + 4n + 2) + n \times (n^3 + 3n^2 + 4n + 2)$
      $= n^6 + 3n^5 + 4n^4 + 2n^3 + n^4 + 3n^3 + 4n^2 + 2n$
      $= n^6 + 3n^5 + 5n^4 + 5n^3 + 4n^2 + 2n$
      [3 marks available — 1 mark for finding an expression for the $(n + 1)$th term, 1 mark for correctly multiplying the expressions for the nth and $(n + 1)$th terms, 1 mark for simplifying to the correct final answer]

8   a) $u_1 = 3$, $u_2 = 6$, $u_3 = 12$, $u_4 = 24$
      [2 marks available — 2 marks for all three values correct, otherwise 1 mark for one or two values correct]
  b)   Each term is 2 × the previous term, so the sequence is exponential. The first term is 3 and you multiply by 2 each time, so the nth term is $3 \times 2^{(n-1)}$.
      [2 marks available — 2 marks for the correct answer, otherwise 1 mark for an expression involving an exponential term]

9   a)   Number of grey squares as a sequence: 1, 5, 9, 13, ...
      Common difference = 4 so $4n$ is in the formula.
      To get from $4n$ to each term, you have to subtract 3, so the expression for the nth term is $4n - 3$.
      [2 marks available — 2 marks for correct expression, otherwise 1 mark for finding $4n$]
  b)   Assume Giles makes the nth and $(n + 1)$th patterns.
      He uses $4n - 3$ grey squares in the nth pattern and $4(n + 1) - 3 = 4n + 4 - 3 = 4n + 1$ grey squares in the $(n + 1)$th pattern *[1 mark]*. He uses 414 grey squares in total, so
      $(4n - 3) + (4n + 1) = 414$ *[1 mark]*
      $8n - 2 = 414$
      $8n = 416$
      $n = 52$
      So Giles has made the 52nd and 53rd patterns *[1 mark]*.
      [3 marks available in total — as above]
  c)   Total number of squares:
                        1   7   17   31
      First difference: 6   10   14
      Second difference: 4   4 *[1 mark]*
      The second differences are constant so the sequence is quadratic. Coefficient of $n^2 = 4 \div 2 = 2$ *[1 mark]*.
      Actual sequence – $2n^2$ sequence:
              –1   –1   –1   –1
      So the nth term of the sequence is $2n^2 - 1$ *[1 mark]*.
      [3 marks available in total — as above]

Page 51: Algebraic Proportion

1 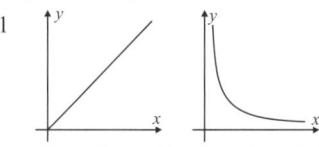

[2 marks available — 1 mark for each correct graph]

Answers

2 a) $a \propto \frac{1}{b}$, so $a = \frac{k}{b}$ *[1 mark]*
   When $a = 436$ and $b = 18$, $436 = \frac{k}{18}$, $k = 7848$ *[1 mark]*
   So $a = \frac{7848}{b}$ *[1 mark]*
   [3 marks available in total — as above]

   b) $327 = \frac{7848}{b}$, so $b = 24$ *[1 mark]*

3 a) $E \propto v^2$, so $E = kv^2$ *[1 mark]*
   When $E = 1008$ and $v = 24$, $1008 = k \times 24^2$,
   so $k = 1008 \div 24^2 = 1.75$ *[1 mark]*. So $E = 1.75v^2$.
   When $v = 32$, $E = 1.75 \times 32^2 = 1792$ *[1 mark]*
   [3 marks available in total — as above]

   b) If E_0 is the original kinetic energy, v_0 is the original velocity
   and E_1 is the kinetic energy after the change, then:
   $E_0 = k(v_0)^2$, so $E_1 = k(1.3v_0)^2$ *[1 mark]*
   $= 1.69k(v_0)^2$ *[1 mark]* $= 1.69E_0$ *[1 mark]*
   So the kinetic energy will increase by 69%. *[1 mark]*
   [4 marks available in total — as above]

Section Three — Graphs, Functions and Calculus

Page 52: Coordinates

1 a) (2, 1) *[1 mark]*

   b)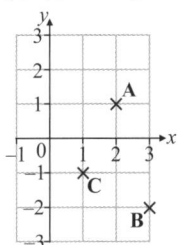
   [1 mark]

   c) (0, 1) *[1 mark]*
   If point D was at this point it would be a scalene triangle, as all the sides would have different lengths.

2 a) $\left(\frac{1+3}{2}, \frac{3+(-1)}{2}\right) = (2, 1)$
   [2 marks available — 1 mark for a correct method, 1 mark for the correct final answer]
   A correct method here is to find the averages of the x- and y-coordinates. Or, you could identify the midpoint of AB on the graph to get your answer — but the first way is much safer.

   b) Comparing coordinates of point **C** and midpoint of **CD**:
   x-distance = 2 – 0 = 2
   y-distance = 1 – –1 = 2
   So to get from the midpoint to point **D**, move up 2 and right 2.
   So point **D** is (2 + 2, 1 + 2) = (4, 3)
   [2 marks available — 1 mark for each correct coordinate]

Pages 53-55: Straight-Line Graphs

1 a)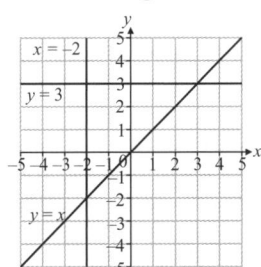
   [3 marks available — 1 mark for each correct line]

   b) (3, 3) *[1 mark]*

2 a)
x	–2	–1	0	1	2
y	–8	–5	–2	1	4

   [2 marks available — 2 marks for all values correct, otherwise 1 mark for 2 correct values]

   b)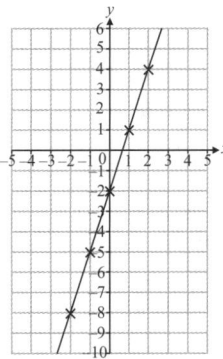
   [2 marks available — 2 marks for a straight line drawn from (–2, –8) to (2, 4), otherwise 1 mark for a straight line with the correct gradient, or a straight line with a positive gradient going through any correctly plotted point]

   c)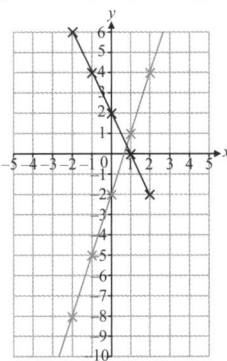
   [3 marks available — 3 marks for a correct line drawn from (–2, 6) to (2, –2), otherwise 2 marks for a line that passes through (0, 2) and has a gradient of –2, or 1 mark for a line passing through (0, 2), or a line with a gradient of –2]

3 a) –2 *[1 mark]*   b) 1 *[1 mark]*   c) $\frac{1}{2}$ *[1 mark]*

4 Find the gradient: $\frac{\text{change in } y}{\text{change in } x} = \frac{4-1}{1-0} = 3$
   Line crosses y-axis at 1, so equation of line is $y = 3x + 1$.
   [3 marks available — 3 marks for a fully correct answer, otherwise 2 marks for a correct gradient, or 1 mark for a correct method to find the gradient]

5 a)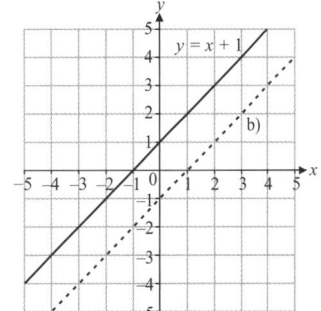
   [2 marks available — 1 mark for the correct gradient, 1 mark for the correct intersection with the y-axis]

   b) Draw line parallel to $y = x + 1$ that passes through (2, 1), — see dashed line on grid above.
   m = 1 and c = –1, so $y = x – 1$
   [2 marks available — 1 mark for the correct line on the graph, 1 mark for the correct equation]

6 The lines are parallel, so their gradients are equal: m = 4 *[1 mark]*
   When $x = –1$, $y = 0$, so put this into $y = 4x + c$
   $0 = (4 \times –1) + c$, so $c = 4$ *[1 mark]*
   So equation of line is $y = 4x + 4$ *[1 mark]*
   [3 marks available in total — as above]

7 $3x + 4 = 2x + 6$ *[1 mark]*
   $x = 2$ *[1 mark]*
   so $y = 3(2) + 4 = 10$ and point *M* is (2, 10) *[1 mark]*
   Line **N** is perpendicular to $y = \frac{1}{2}x + 6$, which has a gradient of $\frac{1}{2}$.
   Gradient of line **N** = $-1 \div \frac{1}{2} = -2$, so $y = -2x + c$ *[1 mark]*
   $10 = -2 \times 2 + c$, so $c = 10 + 4 = 14$
   $y = -2x + 14$ *[1 mark]*
   [5 marks available in total — as above]

8 a) $2a + 4 = 2c$, so $a + 2 = c$
      Substitute values $a + 2 = c$ and $b - 6 = d$ into point (c, d):
      $(c, d) = (a + 2, b - 6)$
      Gradient of **S** = $\frac{b - 6 - b}{a + 2 - a} = \frac{-6}{2} = -3$
      [3 marks available — 1 mark for correctly substituting values into a point, 1 mark for finding change in y over change in x, 1 mark for the correct answer]

   b) Gradient = $\frac{1}{3}$ *[1 mark]*
      So $y = \frac{1}{3}x + c$.
      Substitute (6, 3) into the equation:
      $3 = \frac{1}{3} \times 6 + c$
      $c = 1$ *[1 mark]*
      Line **R**: $y = \frac{1}{3}x + 1$ *[1 mark]*
      [3 marks available in total — as above]

Pages 56-57: Quadratic Graphs

1 a) 0 *[1 mark]*
   b) −2 and 0 *[1 mark]*
   c) (−1, −1) *[1 mark]*
   d) $x = -1$ *[1 mark]*

2 a)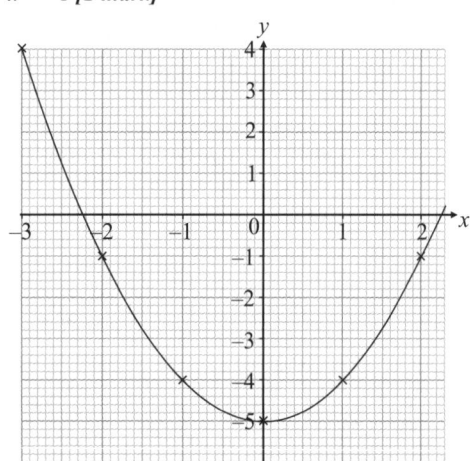

   [3 marks available — 3 marks for a correct smooth curve with all points plotted correctly, otherwise 2 marks if 5 or 6 points are plotted correctly, or 1 mark if 3 or 4 points are plotted correctly]

   b) $x = -2.2$ (allow −2.3 or −2.1) *[1 mark]*

3 a) (1.5, −0.25) *[1 mark]*
      If your y-coordinate is between −0.28 and −0.23 you'll get the mark.
   b) $a = 2$ *[1 mark]*

4 a)
x	−2	−1	0	1	2	3	4
y	−7	−2	1	2	1	−2	−7

   [2 marks available — 1 mark for each correct value]

   b)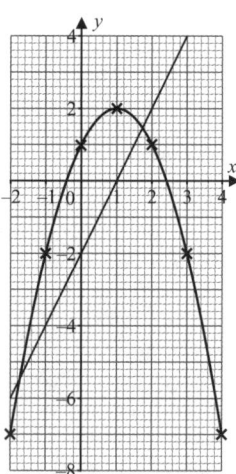

   [3 marks available — 3 marks for a correct smooth curve with all points plotted correctly, otherwise 2 marks if 6 or 7 points are plotted correctly, or 1 mark if 4 or 5 points are plotted correctly]

   c) (i) Find −3 on the *y*-axis and go across until you hit the curve:
         $x = -1.2$ (allow −1.3 or −1.1) *[1 mark]*
         and $x = 3.2$ (allow 3.1 or 3.3) *[1 mark]*
         [2 marks available in total — as above]

      (ii) The solutions are where your quadratic graph intersects the line $y = 2x - 2$:
         $x = -1.7$ (allow −1.8 or −1.6) *[1 mark]*
         and $x = 1.7$ (allow 1.6 or 1.8) *[1 mark]*
         [2 marks available in total — as above]

Pages 58-59: Harder Graphs

1 a)
x	1	2	4	8	10	16
y	8	4	2	1	0.8	0.5

   [2 marks available — 1 mark for each correct value]

   b)

   [3 marks available — 3 marks for a correct smooth curve with all points plotted correctly, otherwise 2 marks if 5 or 6 points are plotted correctly, or 1 mark if 3 or 4 points are plotted correctly]

   c) See above for the correctly plotted line. *[1 mark for a line with the correct gradient, 1 mark for the correct y-intercept]*
      The solution is where the graphs intersect,
      so $x = 1.7$ (allow 1.5, 1.6, 1.8 or 1.9) *[1 mark]*
      [3 marks available in total — as above]

2 a) $y = \frac{1}{x^2}$ *[1 mark]*
   b) $y = x^3 - 3x$ *[1 mark]*
   c) $y = 3^x$ *[1 mark]*

3 a)
x	...	3	3.5	4
y	...	−5	−2.125	4

   [2 marks available — 1 mark for each correct value]

b)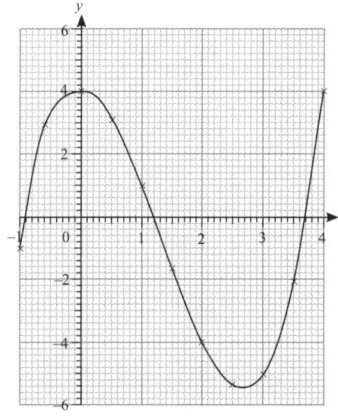

[4 marks available — 4 marks for a correct smooth curve with all points plotted correctly, otherwise 3 marks if 10 or 11 points are plotted correctly, 2 marks if 8 or 9 points are plotted correctly, or 1 mark if 6 or 7 points are plotted correctly]

c) Reading off the graph, where the curve intersects the x-axis,
$x = -0.9$ (allow -1.0 or -0.8) *[1 mark]*
$x = 1.2$ (allow 1.1 or 1.3) *[1 mark]*
$x = 3.7$ (allow 3.6 or 3.8) *[1 mark]*
[3 marks available in total — as above]

4 a)
x	0	1	2	3	4	5
y	11	12	14	18	26	42

[1 mark for each correct value in the table]

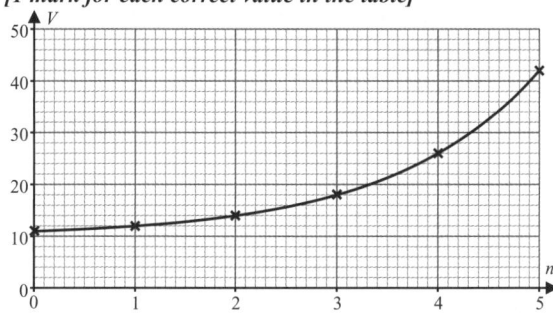

[3 marks for a correct smooth curve with all points plotted correctly, otherwise 2 marks if 5 or 6 points are plotted correctly, or 1 mark if 3 or 4 points are plotted correctly]
[5 marks available in total — as above]

b) Read across from 30 on the vertical axis until you hit the curve: 4.3 years (allow 4.2 years or 4.4 years) *[1 mark]*

Page 60: Real-Life Graphs

1 a)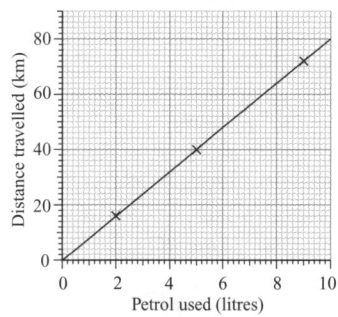

[2 marks available — 1 mark for all points plotted correctly, 1 mark for a straight line joining points]

b) Read along from 28 km and down to 3.5 litres (allow 3.4 to 3.6) *[1 mark]*

c) Gradient = $\frac{\text{change in } y}{\text{change in } x} = \frac{80 - 0}{10 - 0} = 8$

[2 marks available — 1 mark for a correct method to find the gradient, 1 mark for the correct answer]

d) Distance travelled in km per litre of petrol used. *[1 mark]*

2 Graph A and 2, Graph B and 3
Graph C and 4, Graph D and 1
[2 marks available — 2 marks for all four correct pairs, otherwise 1 mark for two correct pairs]

Pages 61-62: Travel Graphs

1 a) Speed = gradient = $6 \div 1 = 6$ km/h *[1 mark]*
b) 2.5 hours *[1 mark]*
c)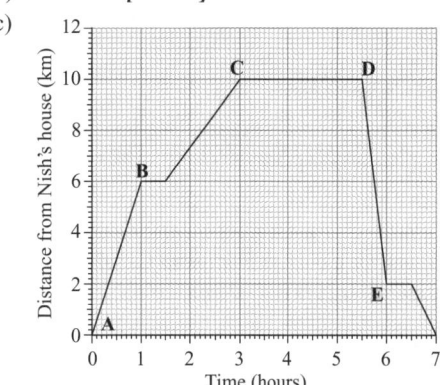

[2 marks available — 1 mark for a flat line from point E for 30 minutes, 1 mark for a straight line from this point to (7, 0)]

2 a)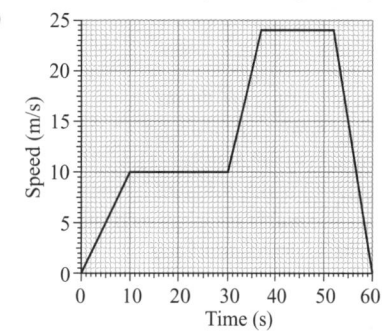

[3 marks available — 3 marks for a fully correct graph, otherwise 2 marks for 2 non-horizontal sections correct, or 1 mark for 1 non-horizontal section correct]

b) At 35 seconds the graph has a gradient of $\frac{24-10}{37-30} = \frac{14}{7} = 2$.
Acceleration = 2 m/s². *[1 mark]*

3 a) The area under the graph is a trapezium where the parallel sides have lengths of 5 and $4 - 1.5 = 2.5$, and the height is 18.
$\frac{1}{2}(5 + 2.5) \times 18$ *[1 mark]* = 67.5, so the total distance travelled by the ferry is 67.5 km. *[1 mark]*
[2 marks available in total — as above]
Alternatively you could split up the area under the graph into different shapes, e.g. 2 triangles and a rectangle.

b) Average speed = $\frac{67.5}{5} = 13.5$ km/h *[1 mark]*

4 a)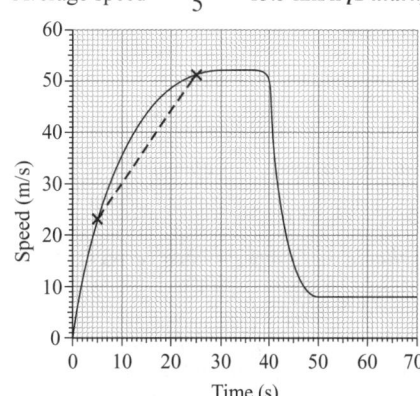

Gradient = $\frac{51 - 23}{25 - 5} = \frac{28}{20} = 1.4$ m/s²

[2 marks available — 2 marks for the correct answer, otherwise 1 mark for reading the speed correctly at 5 seconds and 25 seconds]

b)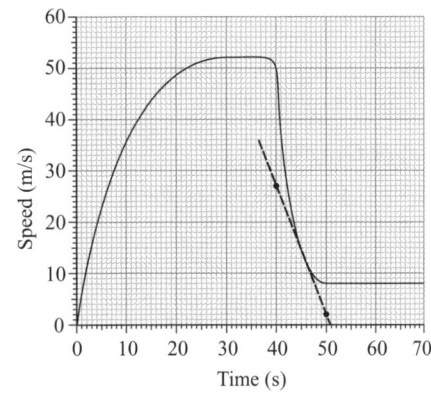

Gradient = $\frac{27-2}{40-50} = \frac{25}{-10} = -2.5$ m/s^2

[3 marks available — 1 mark for drawing the tangent correctly, 1 mark for a correct method to find the gradient, 1 mark for an answer between –2 and –3]

Page 63: Functions

1. a) $\frac{(6 \times 5) - 5}{2} = 12.5$ *[1 mark]*

   b) Let $x = h(y) = \frac{6y - 5}{2}$ *[1 mark]*
   then rearrange to make y the subject:
   $2x = 6y - 5$
   $2x + 5 = 6y$
   $\frac{2x + 5}{6} = y$ so $h^{-1}(x) = \frac{2x + 5}{6}$ *[1 mark]*
   [2 marks available in total — as above]

   c) $h(2) = 3.5$, $h(3) = 6.5$, $h(4) = 9.5$, $h(5) = 12.5$
   So the range of $h(x) = \{3.5, 6.5, 9.5, 12.5\}$
   [2 marks available — 2 marks for all correct numbers, otherwise 1 mark for 2 correct numbers]

2. a) $g(21) = \sqrt{2 \times 21 - 6} = \sqrt{36} = 6$ *[1 mark]*

   b) $gf(x) = g(f(x)) = \sqrt{2(2x^2 + 3) - 6}$ *[1 mark]*
   $= \sqrt{4x^2 + 6 - 6}$
   $= \sqrt{4x^2}$
   $= 2x$ *[1 mark]*
   [2 marks available in total — as above]

   c) $fg(a) = f(g(a)) = 2(\sqrt{2a - 6})^2 + 3$ *[1 mark]*
   $= 2(2a - 6) + 3$
   $= 4a - 9$ *[1 mark]*
   So when $fg(a) = 7$, $4a - 9 = 7$, so $a = 4$ *[1 mark]*
   [3 marks available in total — as above]

Pages 64-65: Differentiation

1. $\frac{dy}{dx} = 6x^2 + 8x$
   [2 marks available — 1 mark for each correct term]

2. a) $\frac{dy}{dx} = -2x$ *[1 mark]*

   b) At $x = -1$: $\frac{dy}{dx} = -2(-1) = 2$ *[1 mark]*
   At $x = 2$: $\frac{dy}{dx} = -2(2) = -4$ *[1 mark]*
   [2 marks available in total — as above]

3. a) $\frac{dy}{dx} = 6x - 3$
   [2 marks available — 1 mark for each correct term]

   b) Turning point is when $\frac{dy}{dx} = 0$.
   $6x - 3 = 0$ *[1 mark]*, so $x = \frac{3}{6} = 0.5$ *[1 mark]*
   Substitute back into the equation to find the y-coordinate:
   $y = (3 \times 0.5^2) - (3 \times 0.5) + 4 = 3.25$, so the coordinates of the turning point are $(0.5, 3.25)$ *[1 mark]*
   [3 marks available in total — as above]

4. a) $y = -x^3 - \frac{21}{2}x^2 + 24x$
   $\frac{dy}{dx} = -3x^2 - 21x + 24$ *[2 marks, otherwise 1 mark for differentiating 2 terms correctly]*
   Turning points are when $\frac{dy}{dx} = 0$.
   $-3x^2 - 21x + 24 = 0$ *[1 mark]*
   $-3(x^2 + 7x - 8) = 0$
   $-3(x + 8)(x - 1) = 0$ *[1 mark]*
   So $x = -8$ and $x = 1$ *[1 mark]*.
   When $x = -8$, $y = -(-8)^3 - \frac{21}{2} \times (-8)^2 + 24 \times -8 = -352$
   When $x = 1$, $y = -(1)^3 - \frac{21}{2} \times 1^2 + 24 \times 1 = 12.5$
   So the turning points are at $(-8, -352)$ and $(1, 12.5)$ *[1 mark]*.
   [6 marks available in total — as above]

   b) $\frac{d^2y}{dx^2} = -6x - 21$
   When $x = -8$, $\frac{d^2y}{dx^2} = (-6 \times -8) - 21 = 27 > 0$ so it's a minimum.
   When $x = 1$, $\frac{d^2y}{dx^2} = (-6 \times 1) - 21 = -27 < 0$ so it's a maximum.
   [3 marks available — 1 mark for any correct method, 1 mark for each correct answer with a reason]

5. $v = -2t^3 + 9t^2 + 39$
   $\frac{dv}{dt} = -6t^2 + 18t$ *[1 mark for each correct term]*
   When $\frac{dv}{dt} = 0$, $-6t^2 + 18t = 0$ *[1 mark]*
   $-6t(t - 3) = 0$ *[1 mark]*
   So $t = 0$ or $t = 3$ *[1 mark]*
   At $t = 0$, $v = 39$ m/s
   At $t = 3$, $v = -2 \times 3^3 + 9 \times 3^2 + 39 = 66$ m/s
   So the maximum velocity is 66 m/s. *[1 mark]*
   [6 marks available in total — as above]

Section Four — Geometry and Measures

Pages 66-67: Geometry

1. Angles on a straight line add up to 180°,
   so $x = 180° - 50° - 30° = 100°$ *[1 mark]*

2. $110° + 170° + 50° + 40° = 370°$. *[1 mark]*
   Angles round a point add up to 360°, not 370°, so these angles do not fit round a point as they are shown on the diagram. *[1 mark]*
   [2 marks available in total — as above]

3. Angle $CBE = 180° - 115° = 65°$
   Angle $BED = 180° - 103° = 77°$ *[1 mark for both]*
   $x + 90° + 77° + 65° = 360°$ *[1 mark]*
   $x + 232° = 360°$
   $x = 360° - 232° = 128°$ *[1 mark]*
   [3 marks available in total — as above]

4. $180° - 48° = 132°$ = Angles $ACB + BAC$ *[1 mark]*
   (angles in a triangle add up to 180°)
   Angle $ACB = 132° \div 2 = 66°$ *[1 mark]* (ABC is isosceles)
   Angle $BCD = 180° - 66° = 114°$ *[1 mark]*
   (angles on a straight line add up to 180°)
   [3 marks available in total — as above]

5. Angles on a straight line add up to 180°,
   so angle $ABJ = 180° - 140° = 40°$ *[1 mark]*
   Co-interior angles add up to 180°,
   so angle $JAB = 180° - 150° = 30°$ *[1 mark]*
   Angles in a triangle add up to 180°,
   so angle $AJB = 180° - 40° - 30° = 110°$ *[1 mark]*
   Angles on a straight line add up to 180°,
   so angle $x = 180° - 110° = 70°$ *[1 mark]*
   [4 marks available in total — as above]

6. $3x + x + 5x + 90 + 81 = 360$
   $9x + 171 = 360$ *[1 mark]*
   $9x = 189$, so $x = 21$ *[1 mark]*
   Angle $ZXY = 180° - x° - 5x° = 180° - 6x°$ *[1 mark]*
   $= 180° - 126° = 54°$ *[1 mark]*
   [4 marks available in total — as above]

7. Angle $FEG = 180° – 70° = 110°$
(angles on a straight line add up to 180°)
Angle $EGH = 145°$ (corresponding angles)
Angle $EGF = 180° – 145° = 35°$
(angles on a straight line add up to 180°)
Angle $EFG = 180° – 35° – 110° = 35°$
(angles in a triangle add up to 180°)
So the triangle must be isosceles as it has two equal angles.
[3 marks available — 3 marks for showing that the triangle is isosceles by finding the two equal angles, otherwise 1 mark for finding angle FEG, 1 mark for finding either angle EGF or angle EFG]

Pages 68-69: Bearings and Scale Drawings

1. a) Drawing of dining table is 4 cm long.
   So 4 cm is equivalent to 2 m.
   $2 ÷ 4 = 0.5$
   Therefore scale is 1 cm to 0.5 m *[1 mark]*
   b) On drawing, dining table is 3 cm from shelves.
   So real distance = $3 × 0.5 = 1.5$ m *[1 mark]*
   c) The chair and the space around it would measure 4 cm × 5 cm on the diagram and there are no spaces that big, so no, it would not be possible.
   [2 marks available — 1 mark for correct answer, 1 mark for reasoning referencing diagram or size of gaps available]

2. Using the scale 1 cm = 100 m:
   400 m = 4 cm and 500 m = 5 cm

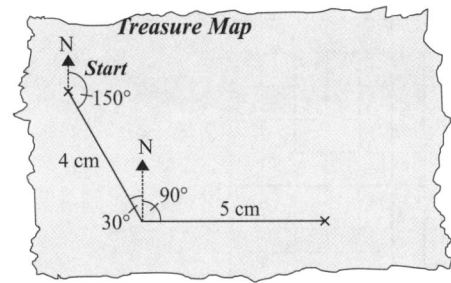

[3 marks available — 1 mark for line on accurate bearing of 150°, 1 mark for line on accurate bearing of 090°, 1 mark for accurate 4 cm and 5 cm line lengths]

3. a)

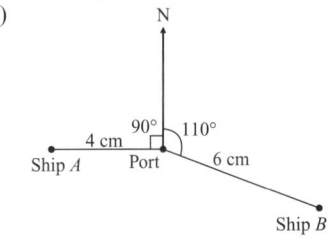

[3 marks available — 1 mark for the correct bearing for Ship A, 1 mark for the correct bearing for Ship B, 1 mark for both ships drawn at the correct distance from Port]
This diagram has been drawn a bit smaller to make it fit — your measurements should match the labels given on the diagram here.

   b) 102° (accept answers between 100° and 104°) *[1 mark]*
   c) $180° – 102° = 78°$
   $360° – 78° = 282°$ (accept answers between 280° and 284°)
   [2 marks available — 1 mark for correctly using 102°, 1 mark for the correct answer]
   You could also do this by adding 180° to 102°.

4. a)

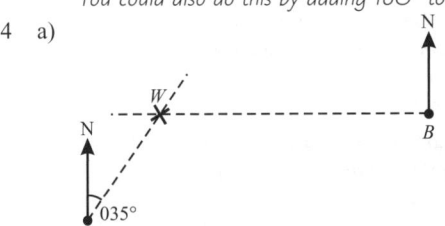

[2 marks available — 1 mark for the correct bearing of 035° from A, 1 mark for marking W directly west of B]

   b)

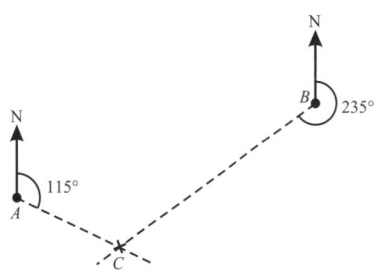

[3 marks available — 1 mark for the correct bearing from A, 1 mark for the correct bearing from B, 1 mark for correctly identifying intersection at point C]

   c) 164° (accept answers between 162° and 166°) *[1 mark]*

Pages 70-71: Polygons and Symmetry

1. a) None *[1 mark]*
   b) Order 2 *[1 mark]*

2. Size of each exterior angle of a regular pentagon:
   $360° ÷ 5$ *[1 mark]* $= 72°$ *[1 mark]*
   [2 marks available in total — as above]

3. a) Rhombus (two lines of symmetry) *[1 mark]*
   b) Three (isosceles triangle, kite and rhombus) *[1 mark]*

4. Angles in a triangle add up to 180°, so the angles in the hexagon add up to $4 × 180° = 720°$ *[1 mark]*

5. a) x is the same as an exterior angle, so $x = 360° ÷ 8$ *[1 mark]*
   $x = 45°$ *[1 mark]*
   [2 marks available in total — as above]
   b) The shape is regular, so angle y is the same as angle OAB.
   Triangle OAB is isosceles, so $y = (180° – 45°) ÷ 2$ *[1 mark]*
   $y = 67.5°$ *[1 mark]*
   [2 marks available in total — as above]

6. Interior angle of regular n-sided polygon = 180° – exterior angle
   $= 180° – (360° ÷ n)$
   Interior angle of regular octagon = $180° – (360° ÷ 8) = 135°$
   Interior angle of regular hexagon = $180° – (360° ÷ 6) = 120°$
   Angle CBK = angle ABC – angle IJK = $135° – 120° = 15°$
   [3 marks available — 1 mark for using a correct method to find an interior angle, 1 mark for finding both interior angles correctly, 1 mark for the correct answer]

7. The polygon has 7 sides, so the interior angles add up to $(7 – 2) × 180° = 900°$ *[1 mark]*
   Using the symmetry of the shape, the angles are:
   120°, 75°, x, 70°, 70°, x and 75° *[1 mark]*
   So $120° + 75° + x + 70° + 70° + x + 75° = 900°$ *[1 mark]*
   $410° + 2x = 900°$, $2x = 490°$, $x = 245°$ *[1 mark]*
   [4 marks available in total — as above]
   You could also use the pentagon created by the dashed line, with angles of 60°, 75°, x, 70° and 90°.

Pages 72-74: Circle Geometry

1. Angle $OBA = 90°$ (tangent meets a radius at 90°) *[1 mark]*
   Angle $OAB = 180° – 90° – 62° = 28°$ (angles in a triangle) *[1 mark]*
   [2 marks available in total — as above]

2. Angle $ACB = 90°$ (the angle in a semicircle is 90°) *[1 mark]*
   Triangle ABC is isosceles, so angle CAB = angle CBA,
   so angle $CBA = (180° – 90°) ÷ 2 = 90° ÷ 2 = 45°$ *[1 mark]*
   [2 marks available in total — as above]

3. a) Angle $OXC = 90°$ as the angle in a semicircle is 90°. *[1 mark]*
   The line OX is a radius of the larger circle, and meets the line CD at 90°, so CD is a tangent to the larger circle at X. *[1 mark]*
   [2 marks available in total — as above]
   b) Angle $AXO = 25°$ *[1 mark]* (AOX is an isosceles triangle)
   Angle $AXC = 25° + 90° = 115°$ *[1 mark]* (from part a)
   Angle $ACX = 180° – 25° – 115° = 40°$ *[1 mark]*
   (angles in a triangle add up to 180°)
   [3 marks available in total — as above]

4 a) Angle *DCO* = 90° *[1 mark]* (tangent meets a radius at 90°)
      Angle *DOC* = 180° − 90° − 24° = 66° *[1 mark]*
      (angles in a triangle)
      [2 marks available in total — as above]
   b) Angle *AOC* = 66° × 2 = 132° *[1 mark]* (tangents from the same point are the same length, so create two identical triangles)
      Angle *ABC* = 132° ÷ 2 = 66° *[1 mark]* (angle at the centre is twice the angle at the circumference)
      Angle *CBE* = 180° − 66° = 114° *[1 mark]*
      (angles on a straight line)
      [3 marks available in total — as above]

5 Angle *FBD* = angle *BCD* = 102°
   (alternate segment theorem) *[1 mark]*
   Angle *CDB* = 180° − 147° = 33° (angles on a straight line) *[1 mark]*
   Angle *CBD* = 180° − 102° − 33° = 45° (angles in a triangle) *[1 mark]*
   Angle *CAD* = angle *CBD* = 45°
   (angles in the same segment are equal) *[1 mark]*
   [4 marks available in total — as above]

6 Opposite angles in a cyclic quadrilateral add up to 180°, so
   angle *ADC* = 180° − angle *ABC* = 180° − 119° = 61°
   Angle *CDX* = angle *ADC* − angle *ADX* = 61° − 31° = 30°
   If *X* was the centre of the circle, *XD* and *XC* would be radii, so triangle *CXD* would be isosceles and angles *CDX* and *XCD* would be equal. Here angle *CDX* = 30° and angle *XCD* = 28° so the angles are not equal, and therefore *X* is not the centre of the circle.
   [3 marks available in total — 1 mark for finding angle ADC, 1 mark for finding angle CDX, 1 mark for using "two radii from an isosceles triangle" to explain why X cannot be the centre]

7 a) Angle *DAB* = Angle *BDE* = 53°
      (alternate segment theorem) *[1 mark]*
      Angle *DOB* = 2 × Angle *DAB* = 106° (angle at the centre is twice the angle at the circumference) *[1 mark]*
      [2 marks available in total — as above]
      You could have done this one by splitting the triangle DOB into two identical right-angled triangles and working out the angles.
   b) *OB* and *OD* are both radii, so *OBD* is an isosceles triangle. The radius *OC* crosses chord *BD* at right angles, so it bisects *BD* *[1 mark]* and divides the isosceles triangle *OBD* in half, which means angle *COB* = 0.5 × angle *DOB* *[1 mark]*.
      [2 marks available in total — as above]

8 *OE* and *OF* pass through the centre of the circle and are perpendicular to the chords *AB* and *CD*, so they bisect the chords, which means *E* and *F* are the midpoints of the chords. *[1 mark]*
   The chords *AB* and *CD* are the same length, so they are the same distance from the centre of the circle, and so are their midpoints. So *OE* is the same length as *OF*. *[1 mark]*
   Triangle *OEF* is isosceles, so angles *OEF* and *OFE* are equal. *[1 mark]*
   Angle *OEF* = Angle *OFE* = (180° − 60°) ÷ 2 = 120° ÷ 2 = 60°
   All three angles in the triangle are 60°, so the triangle is equilateral. *[1 mark]*
   [4 marks available in total — as above]

Page 75: Congruence and Similarity

1 A and H *[1 mark]*
   E and G *[1 mark]*
   [2 marks available in total — as above]

2 Angle *EBD* = 180° − 55° − 65° = 60°
   Angle *x* = angle *EBD* = 60° (vertically opposite angles) *[1 mark]*
   Scale factor from *ABC* to *DBE* = 5 ÷ 2 = 2.5 *[1 mark]*
   So *y* = 6 ÷ 2.5 = 2.4 cm *[1 mark]*
   [3 marks available in total — as above]

3 The piece that has been cut off is similar to the original piece of wood. Scale factor = (2 × 14) ÷ 70 = 28 ÷ 70 = 0.4 *[1 mark]*
   So *x* = 0.4 × (75 + *x*) *[1 mark]*
   x = 0.4 × 75 + 0.4*x*
   0.6*x* = 30
   x = 30 ÷ 0.6 = 50 cm *[1 mark]*
   [3 marks available in total — as above]

Pages 76-77: The Four Transformations

1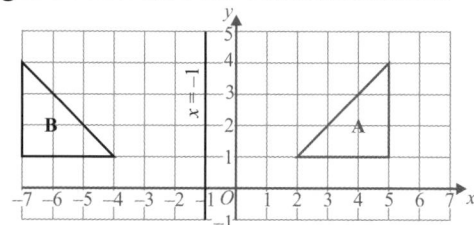
   [2 marks available — 2 marks for the correct reflection, otherwise 1 mark for a triangle reflected but in the wrong position]

2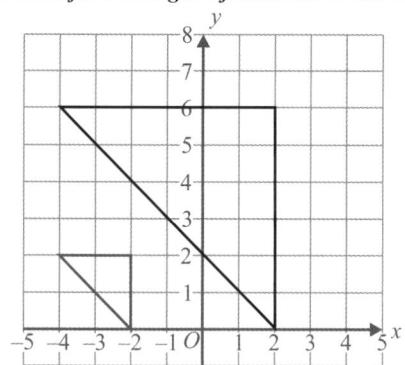
   [2 marks available — 2 marks for the correct enlargement, otherwise 1 mark for an enlargement with either an incorrect scale factor or in the wrong position]

3 a)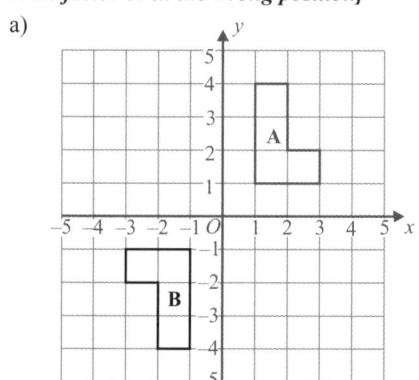
      [2 marks available — 2 marks for a correct rotation, otherwise 1 mark for a rotation of 180° in the wrong position]
   b)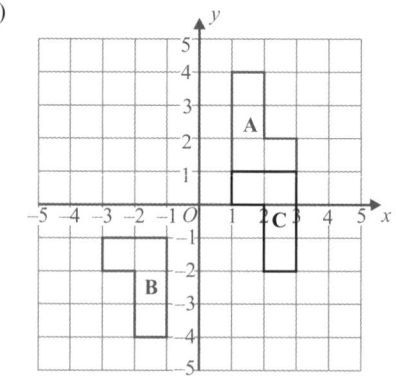
      [2 marks available — 1 mark for a correct translation in the x-direction, 1 mark for a correct translation in the y-direction]
   c) Rotation of 180° about (2, 1).
      [3 marks available — 1 mark for rotation, 1 mark for 180°, 1 mark for the correct centre of rotation]
      You could also say this is an enlargement with scale factor −1 and centre of enlargement (2, 1).

4 a) Reflection *[1 mark]* in the *y*-axis / the line *x* = 0 *[1 mark]*
      [2 marks available in total — as above]

b)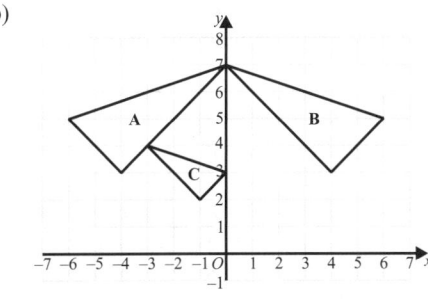

[2 marks available — 2 marks for the correct shape in the correct position, otherwise 1 mark for one coordinate correct]

5 a) Rotation *[1 mark]* by 90° clockwise
(or 270° anticlockwise) *[1 mark]* about (1, 0) *[1 mark]*
[3 marks available in total — as above]

b)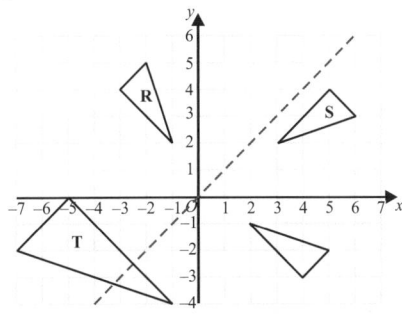

[4 marks available — 2 marks for the correct reflection, otherwise 1 mark for a reflection in the wrong position. 2 marks for the correct enlargement, otherwise 1 mark for an enlargement with either an incorrect scale factor or in the wrong position]

Page 78: More Enlargements

1. $1^3 : 7^3 = 1 : 343$ *[1 mark]*
2. Area of enlarged shape = 7×3^2 *[1 mark]* = 63 cm² *[1 mark]*
[2 marks available in total — as above]
3. Volume of cuboid A = $1.5 \times 12.5 = 18.75$ cm³
Volume of cuboid B = $18.75 \times \left(\frac{8}{5}\right)^3 = 76.8$ cm³
[2 marks available — 1 mark for a correct method, 1 mark for the correct answer]
4. Let x be the scale factor for length from cylinder A to cylinder B.
$x^3 = \frac{64}{27}$, so $x = \sqrt[3]{\frac{64}{27}} = \frac{4}{3}$ *[1 mark]*
Then $x^2 = \frac{4^2}{3^2} = \frac{16}{9}$, so s.a. of cylinder B = $81\pi \times \frac{16}{9}$ *[1 mark]*
= 144π cm² *[1 mark]*
[3 marks available in total — as above]

Page 79: Unit Conversions

1. 1 m² = 100 × 100 = 10 000 cm²
So 39 200 cm² = 39 200 ÷ 10 000 *[1 mark]* = 3.92 m² *[1 mark]*
[2 marks available in total — as above]
2. 150 litres = 150 × 1000 *[1 mark]* = 150 000 cm³
1 m³ = 100 × 100 × 100 = 1 000 000 cm³
So 150 000 cm³ = 150 000 ÷ 1 000 000 *[1 mark]* = 0.15 m³ *[1 mark]*
[3 marks available in total — as above]
3. Convert the cost from the Canadian gas station into US dollars:
1 litre of gasoline costs 1.15 Canadian dollars
= 1.15 × 0.75 = 0.8625 US dollars *[1 mark]*
So a litre of gasoline is 0.8625 − 0.8 = 0.0625 US dollars cheaper in the US than in Canada. *[1 mark]*
40 litres will be 0.0625 × 40 = 2.50 US dollars cheaper *[1 mark]*
[3 marks available in total — as above]
You could also have worked out the cost of 40 litres of gasoline in both countries in US dollars, then subtracted to find the answer.

4. Start by converting the side lengths to the same measurement
1 m = 100 cm = 1000 mm
3 × 1000 = 3000, so 3 m = 3000 mm *[1 mark]*
Work out how many small cubes you could fit along each side of the large cube: 3000 ÷ 60 = 50 *[1 mark]*
So in the large cube you could fit:
50 × 50 × 50 = 125 000 small cubes *[1 mark]*
[3 marks available in total — as above]

Page 80: Time

1. a) She has to wait from 16 58 to 17 04 which is 6 minutes. *[1 mark]*
b) It will take from 16 40 to 18 15:
16 40 to 17 00 is 20 minutes.
17 00 to 18 00 is 1 hour.
18 00 to 18 15 is 15 minutes.
So the total time is 1 hour + 20 minutes + 15 minutes
= 1 hour 35 minutes *[1 mark]*

2. His flight took off when it was
15 50 + 5 hours 30 minutes = 21 20 in India. *[1 mark]*
21 20 to 05 20 is 8 hours
05 20 to 05 45 is 25 minutes
So the flight time was 8 hours 25 minutes *[1 mark]*
[2 marks available in total — as above]

3. a) 9:55 + 2 hours = 11:55
11:55 + 15 minutes = 12:10 *[1 mark]*
The cake is 400 g so she needs to bake it for an extra 10 × 4 = 40 minutes *[1 mark]*
12:10 + 40 minutes = 12:50 *[1 mark]*
[3 marks available in total — as above]
b) 10:45 to 11:00 is 15 minutes
11:00 to 14:00 is 3 hours
14:00 to 14:02 is 2 minutes
So he baked the cake for 3 hours 17 minutes *[1 mark]*
3 hours 17 minutes − 2 hours 15 minutes
= 1 hour 2 minutes = 62 minutes *[1 mark]*
The additional time is one minute per 100 ÷ 10 = 10 g.
So the cake weighed 62 × 10 = 620 g *[1 mark]*
[3 marks available in total — as above]

Pages 81-82: Speed, Density and Pressure

1. 1 hour 15 minutes = 1.25 hours *[1 mark]*
Distance = speed × time, so distance = 56 × 1.25 = 70 km *[1 mark]*
[2 marks available in total — as above]
2. Pressure = force ÷ area, so pressure = 1000 ÷ 50 = 20 N/cm² *[1 mark]*
3. a) Volume = mass ÷ density
Volume of metal A = 120 ÷ 6 = 20 cm³ *[1 mark]*
Volume of metal B = 130 ÷ 5 = 26 cm³ *[1 mark]*
Total volume = 20 + 26 = 46 cm³ *[1 mark]*
[3 marks available in total — as above]
b) Density = mass ÷ volume = (120 + 130) ÷ 46
= 250 ÷ 46 *[1 mark]* = 5.43478... = 5.4 g/cm³ (1 d.p.) *[1 mark]*
[2 marks available in total — as above]
4. Convert the area from m² to km²:
1 km² = 1000 × 1000 = 1 000 000 m²
So 24 000 m² = 24 000 ÷ 1 000 000 *[1 mark]* = 0.024 km² *[1 mark]*
People = area × population density = 0.024 × 8770 = 210.48
= 210 people (to the nearest whole number) *[1 mark]*
[3 marks available in total — as above]
5. 10 cm³ of brass contains 7 cm³ of copper and 3 cm³ of zinc.
7 cm³ of copper has a mass of 7 × 8.9 = 62.3 g
3 cm³ of zinc has a mass of 3 × 7.1 = 21.3 g
10 cm³ of brass has a mass of 62.3 + 21.3 = 83.6 g
Density of brass = 83.6 ÷ 10 = 8.36 g/cm³
[4 marks available — 1 mark for finding the mass of a stated volume of copper or zinc, 1 mark for finding the total mass of a stated volume of brass, 1 mark for attempting to find density using total mass ÷ total volume, 1 mark for the correct final answer]

6 Work out how long the bus journey takes:
Time = distance ÷ speed, so time = 32 ÷ 40 = 0.8 hours
= 48 minutes *[1 mark]*
1 hour 8 minutes = 68 minutes, so she reached the bus stop in:
68 − 5 − 48 = 15 minutes = 0.25 hours *[1 mark]*
Speed = distance ÷ time, so speed = 1.2 ÷ 0.25 = 4.8 km/h *[1 mark]*
[3 marks available in total — as above]

7 a) Area of circular base = $\pi \times (10x)^2 = 100\pi x^2$ cm² *[1 mark]*
$100\pi x^2$ cm² = $(100\pi x^2 \div 100 \div 100)$ m² = $0.01\pi x^2$ m² *[1 mark]*
Weight = $650 \times 0.01\pi x^2 = 6.5\pi x^2$ N *[1 mark]*
[3 marks available in total — as above]

b) E.g. If the diameter is halved, the area of the circular base becomes: $\pi \times (5x)^2 = 25\pi x^2$ cm² = $0.0025\pi x^2$ m²
Pressure = $6.5\pi x^2 \div 0.0025\pi x^2 = 2600$ N/m²
2600 N/m² ÷ 650 N/m² = 4
If the diameter of the circle is halved the pressure increases and is 4 times greater.
[2 marks available — 1 mark for saying that the pressure increases, 1 mark for saying it's 4 times greater]

Pages 83-84: Perimeter and Area

1 Shorter parallel side of the trapezium
= 52 cm − 16 cm − 16 cm = 20 cm
To see this, split the shape into two triangles and a rectangle. The two triangles are both isosceles, so the base of each triangle is 16 cm long.
Area of trapezium = $0.5 \times (52 + 20) \times 16 = 576$ cm²
[2 marks available — 1 mark for finding shorter parallel side of trapezium, 1 mark for correct answer]

2 Area of circular card = $\pi \times 5^2 = 25\pi$ cm² *[1 mark]*
Area of cut out circle = $\pi \times 3^2 = 9\pi$ cm² *[1 mark]*
Area of letter "O" = $25\pi - 9\pi = 16\pi = 50.27$ cm² (2 d.p.) *[1 mark]*
[3 marks available in total — as above]

3 Area of triangle = $0.5 \times 2x \times 2x = 2x^2$ *[1 mark]*
Area of square = x^2
$2x^2 - x^2 = 9$ cm² *[1 mark]*
$x^2 = 9$ cm² so $x = 3$ cm *[1 mark]*
Perimeter of square = 4×3 cm = 12 cm *[1 mark]*
[4 marks available in total — as above]

4 Lawn area = $(30 \text{ m} \times 10 \text{ m}) - (\pi \times (5 \text{ m})^2) = 221.460...$ m²
Boxes of seed needed = $221.460...$ m² $\div 10$ m² = 22.15
So Lynn must buy 23 boxes. Total cost = $23 \times \$7 = \$161$
[3 marks available — 1 mark for correctly calculating the lawn area, 1 mark for dividing the area by 10 m² to find the number of boxes, 1 mark for the correct answer]

5 Each straight section = $2 \times$ radius = $2 \times 9 = 18$ cm *[1 mark]*
Each curved section = $\frac{1}{3} \times$ circumference of circle
$= \frac{1}{3} \times 2 \times \pi \times 9$ *[1 mark]*
= 6π cm *[1 mark]*
Total length = $3 \times 18 + 3 \times 6\pi = 110.5$ cm (1 d.p.) *[1 mark]*
[4 marks available in total — as above]

6 Let A have width x and length y.
Then B has width x and length $2y$.
The perimeter of C is $y + x + y + x + 2y + 2x = 4x + 4y$
and the perimeter of D is $x + 2y + x + 2y + (y - x) + x + y = 2x + 6y$
So, $4x + 4y = 28$ (1)
$2x + 6y = 34$ (2) $\xrightarrow{\times 2}$ $4x + 12y = 68$ (3)
(3) − (1): $8y = 40$, so $y = 5$ cm
Substitute into (1): $4x + 20 = 28$, so $x = 2$ cm
Perimeter of $A = 2 + 5 + 2 + 5 = 14$ cm
Perimeter of $B = 2 + 10 + 2 + 10 = 24$ cm
[6 marks available — 1 mark for setting up simultaneous equations, 1 mark for a correct method for finding one variable, 1 mark for a correct method for finding the other variable, 1 mark for using these values to find perimeters of A and B, 1 mark for perimeter of shape A correct, 1 mark for perimeter of shape B correct]

7 Circumference of full circle = $2 \times \pi \times 4 = 8\pi$ cm
Length of arc = $(27 \div 360) \times$ circumference of circle
= $0.075 \times 8\pi = 0.6\pi$ cm
Perimeter of sector = $0.6\pi + 4 + 4 = 9.88$ cm (3 s.f.)
Area of full circle = $\pi \times 4^2 = 16\pi$ cm²
Area of sector = $(27 \div 360) \times$ area of circle
= $0.075 \times 16\pi = 1.2\pi$ cm² = 3.77 cm² (3 s.f.)
[5 marks available — 1 mark for a correct method for calculating the length of the arc, 1 mark for correct arc length, 1 mark for correct perimeter of sector, 1 mark for a correct method for finding the area of the sector, 1 mark for correct area of sector]

Page 85: Triangle Constructions

1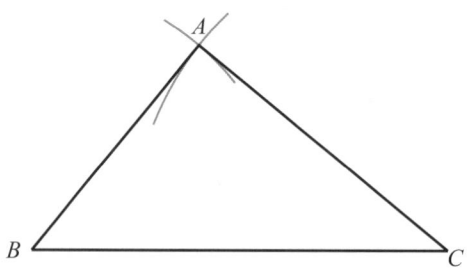
[2 marks available — 2 marks for a correctly drawn triangle, otherwise 1 mark for at least two sides correctly drawn (allow measurements within 1 mm)]

2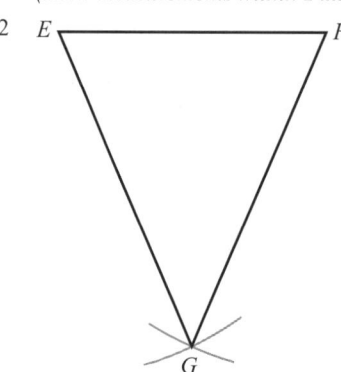
[2 marks available — 1 mark for arcs drawn with a radius of 4.5 cm, 1 mark for completed triangle]

Pages 86-87: 3D Shapes — Surface Area and Nets

1 a)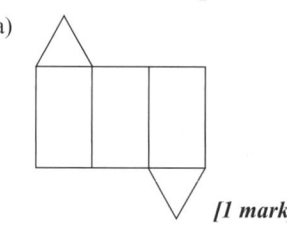
[1 mark]

b) 4 *[1 mark]*

2 The tower has $10 + 10 + 10 + 10 + 1 + 1 = 42$ square faces on its surface. *[1 mark]*
Each square face has an area of $3 \times 3 = 9$ m² *[1 mark]*
so the total surface area = $9 \times 42 = 378$ m² *[1 mark]*
[3 marks available in total — as above]

3 Surface area of sphere = $4\pi r^2 = 4 \times \pi \times 4^2$ *[1 mark]*
= $64\pi = 201.0619...$ cm² = 201.06 cm² (2 d.p.) *[1 mark]*
[2 marks available in total — as above]

4 Area of one face = $0.5 \times 2 \times \sqrt{3} = \sqrt{3}$ cm² *[1 mark]*
So the surface area = $4 \times \sqrt{3} = 6.93$ cm² (2 d.p.) *[1 mark]*
[2 marks available in total — as above]

5 a) Area of triangle = ½ × 6.0 × 5.2 = 15.6 cm² *[1 mark]*
Area of whole octahedron = $8 \times 15.6 = 124.8$ cm² *[1 mark]*
[2 marks available in total — as above]

b) Surface area of octahedron = $8 \times \frac{1}{2} \times b \times h = 4bh$ *[1 mark]*
Surface area of cylinder = $2\pi bh + 2\pi b^2$ *[1 mark]*
$2\pi bh > 4bh$ *[1 mark]* and $2\pi b^2$ is always positive so $2\pi bh + 2\pi b^2 > 4bh$, which means the surface area of the cylinder is greater than that of the octahedron for any values of b and h. *[1 mark for a correct explanation]*
[4 marks available in total — as above]

6 a) Length of arc = $(90 \div 360) \times (2 \times \pi \times 16) = 8\pi$ *[1 mark]*
The circumference of the base = 8π,
so the radius of the base is $8\pi \div 2\pi = 4$ cm *[1 mark]*
[2 marks available in total — as above]

b) Slanting length of cone = 16 cm
Curved surface area of cone = $(\pi \times 4 \times 16)$ *[1 mark]*
Area of base of cone = $(\pi \times 4^2)$ *[1 mark]*
Total surface area of cone = $(\pi \times 4 \times 16) + (\pi \times 4^2)$
= $80\pi = 251.33$ cm^2 (2 d.p.) *[1 mark]*
[3 marks available in total — as above]

Pages 88-89: 3D Shapes — Volume

1 Volume of sphere = $\frac{4}{3}\pi r^3 = \frac{4}{3} \times \pi \times 12^3$ *[1 mark]* = 2304π cm^3
= 7238.22... = 7240 cm^3 (3 s.f.) *[1 mark]*
[2 marks available in total — as above]

2 Area of cross-section = $\frac{1}{2} \times 6 \times 5 = 15$ cm^2 *[1 mark]*
Volume of prism = $15 \times 6 = 90$ cm^3 *[1 mark]*
[2 marks available in total — as above]

3 Volume = $\frac{4}{3}\pi r^3 = 478$ cm^3 *[1 mark]*
$r = \sqrt[3]{\frac{3 \times 478}{4\pi}} = 4.8504...$ cm *[1 mark]*
Surface area = $4\pi r^2 = 4\pi \times (4.8504...)^2$ *[1 mark]*
= 295.6 cm^2 (1 d.p.) *[1 mark]*
[4 marks available in total — as above]

4 Volume of cuboid = $1 \times 3 \times 3 = 9$ m^3 *[1 mark]*
Volume of pyramid = $\frac{1}{3} \times (3 \times 3) \times (2.5 - 1) = 4.5$ m^3 *[1 mark]*
So the total volume = $9 + 4.5 = 13.5$ m^3 *[1 mark]*
[3 marks available in total — as above]

5 Volume of cone = $\frac{1}{3}(\pi \times 6^2 \times 18) = 216\pi$ cm^3 *[1 mark]*
So $\frac{4}{3}\pi r^3 = 216\pi$ cm^3 *[1 mark]*
$r^3 = 162$ *[1 mark]*
$r = 5.4513...$ cm = 5.45 cm (3 s.f.) *[1 mark]*
[4 marks available in total — as above]

6 a) Cross-sectional area of pipe: $0.2^2 \times \pi = 0.12566...$ m^2
Cross-sectional area of water:
$0.12566... \div 2 = 0.06283...$ m^2 *[1 mark]*

b) Rate of flow = 2520 litres per minute
= $2520 \div 60$ litres per second
= 42 litres per second *[1 mark]*
= 42 000 cm^3/s
= 0.042 m^3/s *[1 mark]*
Speed = Rate of flow ÷ cross-sectional area of water
= 0.042 m^3/s ÷ 0.06283... m^2 *[1 mark]*
= 0.66845... m/s = 0.668 m/s (3 s.f.) *[1 mark]*
[4 marks available in total — as above]

Section Five — Pythagoras, Trigonometry and Vectors

Page 90: Pythagoras' Theorem

1 Use Pythagoras' theorem on the triangle formed by one step and the ramp. Let x be the length of the ramp.
$20^2 + 55^2 = x^2$ *[1 mark]*
$x^2 = 3425$, so $x = \sqrt{3425}$ cm *[1 mark]*
Total length of the ramp = $2 \times \sqrt{3425}$ cm
= 117.046... cm = 117 cm (3 s.f.) *[1 mark]*
[3 marks available in total — as above]
You could also combine the two steps into one triangle, which would give you the equation $40^2 + 110^2 = c^2$.

2 Let b be the width of the rectangle. The rectangle is split into two right-angled triangles with sides of length b cm, 3 cm and 5 cm.
$5^2 = 3^2 + b^2$ *[1 mark]*
$b^2 = 25 - 9 = 16$
$b = \sqrt{16} = 4$ cm *[1 mark]*
Area of rectangle = 3 cm $\times$ 4 cm = 12 cm^2 *[1 mark]*
[3 marks available in total — as above]

3 Difference in x-coordinates = $8 - 2 = 6$
Difference in y-coordinates = $8 - -1 = 9$ *[1 mark for both]*
So length of line segment = $\sqrt{6^2 + 9^2} = \sqrt{36 + 81}$ *[1 mark]*
= $\sqrt{117}$ = 10.8166... = 10.82 (2 d.p.) *[1 mark]*
[3 marks available in total — as above]

Pages 91-92: Trigonometry — Sin, Cos and Tan

1 $\cos x = \frac{\text{Adjacent}}{\text{Hypotenuse}} = \frac{4}{5} = 0.8$ *[1 mark]*
$x = \cos^{-1}(0.8) = 36.8698...° = 36.9°$ (1 d.p.) *[1 mark]*
[2 marks available in total — as above]

2 a) (i) $\sin 38° = \frac{3}{AC}$ *[1 mark]*
$AC = \frac{3}{\sin 38°} = 4.8728...$ cm = 4.87 cm (2 d.p.) *[1 mark]*
[2 marks available in total — as above]

(ii) $\tan 38° = \frac{3}{DC}$ *[1 mark]*
$DC = \frac{3}{\tan 38°} = 3.8398...$ cm = 3.84 cm (2 d.p.) *[1 mark]*
[2 marks available in total — as above]

b) Area of triangle ACD = $0.5 \times 3.8398... \times 3 = 5.7597...$ cm^2
Area of triangle ABC = $0.5 \times 4.8728... \times 12 = 29.2368...$ cm^2
[1 mark for the correct area of either triangle]
Area of quadrilateral $ABCD$
= $5.7597... + 29.2368...$
= $34.9965...$ cm^2 = 35 cm^2 (2 s.f.) *[1 mark]*
[2 marks available in total — as above]

3 $\tan 60° = \frac{4}{y}$ *[1 mark]*
$y = \frac{4}{\tan 60°}$ *[1 mark]*
$y = \frac{4}{\sqrt{3}}$ m *[1 mark]*
[3 marks available in total — as above]

4 Angle of depression at S = angle of elevation at R

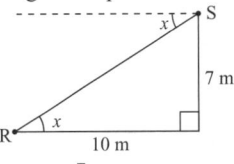

$\tan x = \frac{7}{10}$ *[1 mark]*
$x = \tan^{-1}(0.7) = 34.9920...° = 35.0°$ (1 d.p.) *[1 mark]*
[2 marks available in total — as above]

5 The shortest distance is the perpendicular line from the shoreline to point B.

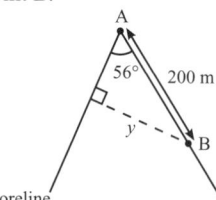

$\sin 56° = \dfrac{y}{200}$ *[1 mark]*

$y = 200 \times \sin 56° = 165.8075...$ m $= 166$ m (3 s.f.) *[1 mark]*

[2 marks available in total — as above]

6 The shortest distance, x, is a perpendicular line from any edge to the centre of the circle. The radius bisects the interior angle, forming angle a.

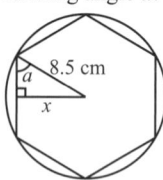

Sum of the interior angles of a hexagon $= 4 \times 180° = 720°$
Each interior angle of a hexagon $= 720° \div 6 = 120°$ *[1 mark]*
$a = 120 \div 2 = 60°$ *[1 mark]*
$\sin 60° = \dfrac{x}{8.5}$ *[1 mark]*
$x = 8.5 \times \sin 60° = 7.3612... = 7.36$ cm (2 d.p.) *[1 mark]*

[4 marks available in total — as above]

You could also say that the other angle in the triangle is $360 \div 12$, then do the calculation $x = \cos 30° \times 8.5$.

Pages 93-94: The Sine and Cosine Rules

1 a) $a^2 = b^2 + c^2 - 2bc \cos A$
   $PQ^2 = 9^2 + 8^2 - (2 \times 9 \times 8 \times \cos 24°)$
   [1 mark for using the correct formula,
   1 mark for substituting the correct numbers]
   $PQ^2 = 13.4494...$ *[1 mark]*
   $PQ = 3.6673...$ cm $= 3.67$ cm *[1 mark]*
   [4 marks available in total — as above]

   b) $\dfrac{a}{\sin A} = \dfrac{b}{\sin B}$
   $\dfrac{3.6673...}{\sin 24°} = \dfrac{8}{\sin QPR}$
   [1 mark for using the correct formula,
   1 mark for substituting the correct numbers]
   $\sin QPR = \dfrac{8 \sin 24°}{3.6673...} = 0.8872...$ *[1 mark]*
   $QPR = \sin^{-1}(0.8872...) = 62.530...° = 62.5°$ (1 d.p.) *[1 mark]*
   [4 marks available in total — as above]

2 *First you need to find one angle using the cosine rule.*
   E.g. use angle CAB.
   $\cos A = \dfrac{b^2 + c^2 - a^2}{2 \times b \times c}$
   $\cos CAB = \dfrac{14^2 + 12^2 - 19^2}{2 \times 14 \times 12}$
   [1 mark for using the correct formula,
   1 mark for substituting the correct numbers]
   $CAB = \cos^{-1}\left(\dfrac{-21}{336}\right) = 93.58...°$ *[1 mark]*
   Area $= \dfrac{1}{2} \times 14 \times 12 \times \sin 93.58...°$ *[1 mark]*
   Area $= 83.835... = 83.84$ cm^2 (2 d.p.) *[1 mark]*
   [5 marks available in total — as above]

3 $\dfrac{a}{\sin A} = \dfrac{b}{\sin B}$
   $\dfrac{36}{\sin ABC} = \dfrac{17}{\sin 26°}$
   [1 mark for using the correct formula,
   1 mark for substituting the correct numbers]
   $\sin ABC = \dfrac{36 \times \sin 26°}{17}$ *[1 mark]* $= 0.92831...$
   $\sin^{-1}(0.92831...) = 68.173...$ Angle ABC is obtuse, so $ABC = 180° - 68.173... = 111.826... = 111.8°$ (1 d.p.) *[1 mark]*
   [4 marks available in total — as above]

4 Area of triangle $ABC = 42$ m^2
   So $\dfrac{1}{2} \times 9 \times AC \times \sin 38° = 42$ *[1 mark]*
   $AC = 15.1598...$ m *[1 mark]*
   $\cos A = \dfrac{b^2 + c^2 - a^2}{2 \times b \times c}$
   $\cos x = \dfrac{12^2 + 11^2 - 15.1598...^2}{2 \times 12 \times 11}$
   [1 mark for using the correct formula,
   1 mark for substituting the correct numbers]
   $x = \cos^{-1}\left(\dfrac{35.1790...}{264}\right) = 82.3423...°$
   $= 82°$ (to the nearest degree) *[1 mark]*
   [5 marks available in total — as above]

5 First, split $ABCD$ into two triangles, ABC and ACD.
   $\dfrac{55}{\sin ACB} = \dfrac{93}{\sin 116°}$ *[1 mark]*
   $\sin ACB = \dfrac{\sin 116°}{93} \times 55$
   Angle $ACB = \sin^{-1}(0.531...) = 32.109...°$ *[1 mark]*
   Angle $BAC = 180° - 116° - 32.109...°$ so,
   Area of $ABC = \dfrac{1}{2} \times 93 \times 55 \times \sin(180 - 116 - 32.10...)°$ *[1 mark]*
   $= 1351.106...$cm^2 *[1 mark]*
   Angle $ACD = 78° - 32.109...°$ so
   Area of $ACD = \dfrac{1}{2} \times 93 \times 84 \times \sin(78 - 32.10...)°$
   $= 2804.531...$cm^2 *[1 mark]*
   Area of $ABCD = 1351.106... + 2804.531...$
   $= 4155.637... = 4160$ cm^2 (3 s.f.) *[1 mark]*
   [6 marks available in total — as above]

Page 95: Trig Graphs

1 a)

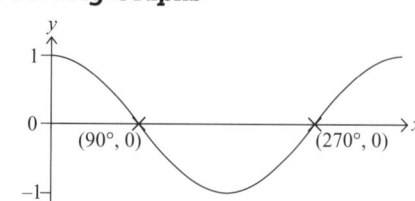

[2 marks available — 1 mark for a curve fluctuating between 1 and –1 on the y-axis, 1 mark for the intersections labelled correctly]

b)

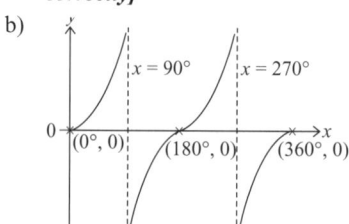

[3 marks available — 1 mark for a curve with the correct shape, 1 mark for the intersections labelled correctly, 1 mark for showing $\tan x$ is undefined at $x = 90°$ and $x = 270°$]

2 $4 \sin x + 1 = 0$
   $\sin x = -\dfrac{1}{4}$ *[1 mark]* so $x = \sin^{-1}\left(-\dfrac{1}{4}\right) = -14.4775...°$ *[1 mark]*
   Now use the graph to work out the solutions for $0° \leq x \leq 360°$:
   $x = 360° - 14.4775...° = 345.5224...° = 345.5°$ (1 d.p.) *[1 mark]*
   $x = 180° + 14.4775...° = 194.4775...° = 194.5°$ (1 d.p.) *[1 mark]*
   [4 marks available in total — as above]

Answers

Page 96: 3D Pythagoras

1. a) $BH^2 = 6^2 + 3^2 + 4^2$ *[1 mark]*
   $BH = \sqrt{61}$ *[1 mark]*
   $BH = 7.81$ cm (2 d.p.) *[1 mark]*
   [3 marks available in total — as above]

   b) Add dimension labels to the diagram:

   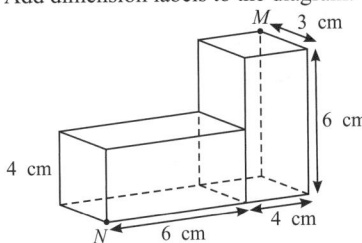

   $MN^2 = (6 + 4)^2 + 6^2 + 3^2$ *[1 mark]*
   $MN = \sqrt{145}$ *[1 mark]*
   $MN = 12.04$ cm (2 d.p.) *[1 mark]*
   [3 marks available in total — as above]

2. Using Pythagoras' theorem on triangle AXV:
   $AX^2 = 8.9^2 - 7.2^2 = 27.37$, so $AX = \sqrt{27.37}$ *[1 mark]*
   and $AC = 2\sqrt{27.37}$ *[1 mark]*
   Now using Pythagoras' theorem on triangle ABC:
   $AB^2 = (2\sqrt{27.37})^2 - 4.2^2 = 91.84$ *[1 mark]*,
   so $AB = \sqrt{91.84} = 9.583...$
   $= 9.58$ cm (3 s.f.) *[1 mark]*
   [4 marks available in total — as above]

Page 97: 3D Trigonometry

1. a) $TY = 8$ cm
   $WY^2 = 12^2 + 5^2 = 169$ *[1 mark]*
   $WY = \sqrt{169} = 13$ cm *[1 mark]*
   Now use the triangle TWY to figure out the angle, x, that TW makes with the base:
   $\tan x = \frac{8}{13}$ *[1 mark]*
   $x = \tan^{-1}\frac{8}{13} = 31.6075...° = 31.6°$ (1 d.p.) *[1 mark]*
   [4 marks available in total — as above]

   b) $WV = 8$ cm
   $VR^2 = 5^2 + (12 \div 2)^2 = 61$ *[1 mark]*
   $VR = \sqrt{61} = 7.8102...$ cm *[1 mark]*
   Now use the triangle WVR to figure out the angle, y, that WR makes with the top:
   $\tan y = \frac{8}{7.8102...}$ *[1 mark]*
   $y = \tan^{-1}\frac{8}{7.8102...} = 45.6876...°$
   $= 46°$ (to the nearest degree) *[1 mark]*
   [4 marks available in total — as above]

2. You can make the following right-angled triangle:

   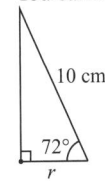

   $\cos 72° = \frac{r}{10}$ *[1 mark]*
   $r = 10 \cos 72° = 3.0901...$ cm *[1 mark]*
   Total surface area $= \pi r l + \pi r^2$
   $= (\pi \times 3.0901... \times 10) + (\pi \times 3.0901...^2)$ *[1 mark]*
   $= 127.0800...$ cm² $= 127$ cm² (3 s.f.) *[1 mark]*
   [4 marks available in total — as above]

Pages 98-99: Vectors

1. a) $\mathbf{a} - \mathbf{b} = \begin{pmatrix}-3\\5\end{pmatrix} - \begin{pmatrix}5\\4\end{pmatrix} = \begin{pmatrix}-8\\1\end{pmatrix}$ *[1 mark]*

   b) $4\mathbf{b} - \mathbf{c} = 4 \times \begin{pmatrix}5\\4\end{pmatrix} - \begin{pmatrix}-4\\-6\end{pmatrix} = \begin{pmatrix}20\\16\end{pmatrix} - \begin{pmatrix}-4\\-6\end{pmatrix} = \begin{pmatrix}24\\22\end{pmatrix}$ *[1 mark]*

   c) $2\mathbf{a} + \mathbf{b} + 3\mathbf{c} = 2 \times \begin{pmatrix}-3\\5\end{pmatrix} + \begin{pmatrix}5\\4\end{pmatrix} + 3 \times \begin{pmatrix}-4\\-6\end{pmatrix}$
   $= \begin{pmatrix}-6\\10\end{pmatrix} + \begin{pmatrix}5\\4\end{pmatrix} + \begin{pmatrix}-12\\-18\end{pmatrix} = \begin{pmatrix}-13\\-4\end{pmatrix}$ *[1 mark]*
   $|2\mathbf{a} + \mathbf{b} + 3\mathbf{c}| = \sqrt{13^2 + 4^2}$ *[1 mark]*
   $= \sqrt{185} = 13.6014... = 13.60$ (2 d.p.) *[1 mark]*
   [3 marks available in total — as above]

2. a) $\vec{AL} = \frac{1}{2} \times \vec{AC} = \frac{1}{2} \times 2\mathbf{c} + \frac{1}{2} \times 2\mathbf{d}$ *[1 mark]* $= \mathbf{c} + \mathbf{d}$ *[1 mark]*
   [2 marks available in total — as above]

   b) $\vec{BL} = \vec{BA} + \vec{AL}$ *[1 mark]* $= -2\mathbf{c} + \mathbf{c} + \mathbf{d} = -\mathbf{c} + \mathbf{d}$ *[1 mark]*
   [2 marks available in total — as above]

   c) $\vec{BC} = 2\mathbf{d} = 2\begin{pmatrix}2\\5\end{pmatrix} = \begin{pmatrix}4\\10\end{pmatrix}$
   $|\vec{BC}| = \sqrt{4^2 + 10^2}$ *[1 mark]*
   $= \sqrt{116} = 10.7703... = 10.77$ (2 d.p.) *[1 mark]*
   [3 marks available in total — as above]

3. a) $\vec{BC} = -\vec{CB} = -(-6\mathbf{a} + 4\mathbf{b}) = 6\mathbf{a} - 4\mathbf{b}$
   $\vec{AC} = \vec{AB} + \vec{BC}$ *[1 mark]*
   $= 3\mathbf{a} + \mathbf{b} + 6\mathbf{a} - 4\mathbf{b} = 9\mathbf{a} - 3\mathbf{b}$ *[1 mark]*
   [2 marks available in total — as above]

   b) $\vec{AP} = \vec{AB} + \frac{1}{2}\vec{BC}$ *[1 mark]*
   $= 3\mathbf{a} + \mathbf{b} + \frac{1}{2} \times (6\mathbf{a} - 4\mathbf{b})$
   $= 6\mathbf{a} - \mathbf{b}$ *[1 mark]*
   [2 marks available in total — as above]

   c) $\vec{CD} = -9\mathbf{a} - 3\mathbf{b} = -3(3\mathbf{a} + \mathbf{b}) = -3 \times \vec{AB}$ *[1 mark]*
   $\vec{CD}$ is a scalar multiple of $\vec{AB}$ so they are parallel. *[1 mark]*
   [2 marks available in total — as above]

4. a) $\vec{OM} = \vec{OA} + \vec{AM} = \vec{OA} + \frac{1}{2}\vec{AB}$ *[1 mark]*
   $\vec{AB} = \mathbf{b} - 2\mathbf{a}$ or $-2\mathbf{a} + \mathbf{b}$
   $\vec{OM} = 2\mathbf{a} + \frac{1}{2}(-2\mathbf{a} + \mathbf{b})$ or $\vec{OM} = 2\mathbf{a} + \frac{1}{2}(\mathbf{b} - 2\mathbf{a})$
   $= \mathbf{a} + \frac{1}{2}\mathbf{b}$ *[1 mark]*
   [2 marks available in total — as above]

   b) $\vec{OX} = \vec{OA} + \vec{AX}$ *[1 mark]*
   As $AX:XB = 1:3$, AX must be one quarter of AB, so:
   $\vec{OX} = \vec{OA} + \frac{1}{4}\vec{AB}$
   $\vec{OX} = 2\mathbf{a} + \frac{1}{4}(\mathbf{b} - 2\mathbf{a})$ *[1 mark]*
   $\vec{OX} = \frac{3}{2}\mathbf{a} + \frac{1}{4}\mathbf{b}$ *[1 mark]*
   [3 marks available in total — as above]

Section Six — Probability and Statistics

Page 100: Probability Basics

1. a) P(not spotty) = 1 − P(spotty) = 1 − 0.25 = 0.75 *[1 mark]*
   P(spotty) means "the probability of choosing a spotty sock".

   b) $3x = 0.75$ so $x = 0.25$ *[1 mark]*
   P(stripy) = $2x = 0.25 \times 2 = 0.5$ *[1 mark]*
   [2 marks available in total — as above]

2. a) P(strawberry) = $\frac{2}{2+5} = \frac{2}{7}$ *[1 mark]*

   b) P(banana) = $\frac{5}{7}$
   $2 \times$ P(strawberry) $= 2 \times \frac{2}{7} = \frac{4}{7}$ so Ami is wrong,
   as she is more than twice as likely to pick a banana sweet.
   [1 mark for saying Ami is wrong with a valid explanation]

3. Number of red counters $= p - n$ *[1 mark]*
   Probability of getting a red counter $= \frac{p-n}{p}$ *[1 mark]*
   [2 marks available in total — as above]

Pages 101-102: Finding Probabilities and Expected Frequency

1. a) EHM, EMH, HME, HEM, MEH, MHE
   [2 marks available — 2 marks for listing all 6 correct combinations, otherwise 1 mark if at least 3 combinations are correct]

   b) There are 6 possible combinations and in 3 of them she does Maths before English (HME, MEH, MHE).
   So P(Maths before English) = $\frac{3}{6} = \frac{1}{2}$ *[1 mark]*

2. a) (Hockey, Netball), (Hockey, Choir), (Hockey, Orienteering), (Orchestra, Netball), (Orchestra, Choir), (Orchestra, Orienteering), (Drama, Netball), (Drama, Choir), (Drama, Orienteering).
   [2 marks available — 2 marks for listing all 9 correct combinations, otherwise 1 mark if at least 5 combinations are correct]

   b) (i) There are 9 combinations and 1 of them is hockey and netball, so P(hockey and netball) = $\frac{1}{9}$ *[1 mark]*

   (ii) There are 9 combinations and 3 of them involve drama on Monday, so P(drama on Monday) = $\frac{3}{9} = \frac{1}{3}$ *[1 mark]*

   You could also count the choices for Monday — there are 3, and 1 of them is drama.

   c) There are 9 combinations, and 3 of them involve choir, so P(choir) = $\frac{3}{9} = \frac{1}{3}$ *[1 mark]*
   Expected frequency = $42 \times \frac{1}{3} = 14$ *[1 mark]*
   [2 marks available in total — as above]

3. a)

	2	4	6	8	10
1	3	5	7	9	11
2	4	6	8	10	12
3	5	7	9	11	13
4	6	8	10	12	14
5	7	9	11	13	15
6	8	10	12	14	16

   [2 marks available — 2 marks if all entries are correct, otherwise 1 mark if at least 4 entries are correct]

   b) There are 30 possible outcomes and 9 of them will score 12 or more. So P(12 or more) = $\frac{9}{30} = \frac{3}{10}$
   [2 marks available — 1 mark for finding that 9 outcomes score 12 or more, 1 mark for the correct answer]

   c) There are 3 outcomes that give a score of 8: 2 and 6, 4 and 4, and 6 and 2. *[1 mark]* Of these, only 1 has the card be a 6, so the probability is $\frac{1}{3}$ *[1 mark]*
   [2 marks available in total — as above]

4. a) $200 \times 0.64 = 128$ times *[1 mark]*

   b) P(Not on time) = $1 - 0.64 = 0.36$ *[1 mark]*
   $300 \times 0.36 = 108$ times *[1 mark]*
   [2 marks available in total — as above]
   You could also estimate that he'll be on time $300 \times 0.64 = 192$ times, and subtract this from 300.

Page 103: The AND/OR Rules

1. a) P(4 or 5) = P(4) + P(5) = $0.25 + 0.1$ *[1 mark]* $= 0.35$ *[1 mark]*
   [2 marks available in total — as above]

   b) P(1 and 3) = P(1) × P(3) = 0.3×0.2 *[1 mark]* $= 0.06$ *[1 mark]*
   [2 marks available in total — as above]

2. P(at least 1 is late) = 1 – P(neither is late)
   P(Alisha isn't late) = $1 - 0.9 = 0.1$
   P(Anton isn't late) = $1 - 0.8 = 0.2$ *[1 mark for both]*
   P(neither is late) = $0.1 \times 0.2 = 0.02$ *[1 mark]*
   P(at least 1 is late) = $1 - 0.02 = 0.98$ *[1 mark]*
   [3 marks available in total — as above]
   You could also solve this question by finding P(exactly 1 is late) and P(both are late) and adding them together:
   (0.1 × 0.8) + (0.9 × 0.2) + (0.8 × 0.9) = 0.98.

3. a) P(losing) = $1 - 0.3 = 0.7$
   P(losing 3 games) = $0.7 \times 0.7 \times 0.7$ *[1 mark]* $= 0.343$ *[1 mark]*
   [2 marks available in total — as above]

   b) P(wins at least one prize) = 1 – P(doesn't win)
   $= 1 - (0.7 \times 0.7)$ *[1 mark]*
   $= 1 - 0.49 = 0.51$ *[1 mark]*
   [2 marks available in total — as above]

   c) P(winning 1 prize in 3 games)
   = P(win, lose, lose) + P(lose, win, lose) + P(lose, lose, win)
   = $(0.3 \times 0.7 \times 0.7) + (0.7 \times 0.3 \times 0.7) + (0.7 \times 0.7 \times 0.3)$ *[1 mark]*
   = $0.147 + 0.147 + 0.147 = 0.441$ *[1 mark]*
   Reza is wrong — there is only a 44.1% chance of winning exactly once in 3 games. *[1 mark]*
   [3 marks available in total — as above]

Pages 104-105: Tree Diagrams

1. a)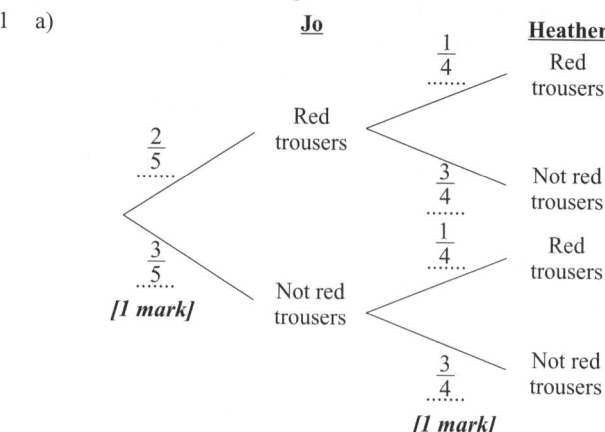
   [2 marks available in total — as above]

   b) P(neither wear red trousers) = $\frac{3}{5} \times \frac{3}{4}$ *[1 mark]* $= \frac{9}{20}$ *[1 mark]*
   [2 marks available in total — as above]

2. a) P(Paul's point) = P(1, 2, 3 or 6) = $\frac{2}{3}$
   P(Jen's point) = $1 - \frac{2}{3} = \frac{1}{3}$ *[1 mark for both probabilities]*
   You can draw a tree diagram to help you:

   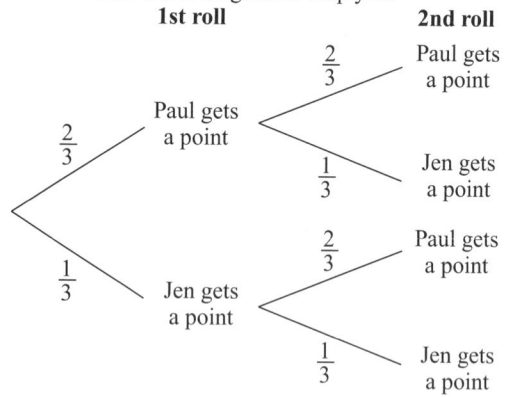

   P(they draw after two rolls) = P(they both get 1 point)
   $= \left(\frac{2}{3} \times \frac{1}{3}\right) + \left(\frac{1}{3} \times \frac{2}{3}\right)$ *[1 mark]*
   $= \frac{2}{9} + \frac{2}{9} = \frac{4}{9}$ *[1 mark]*
   [3 marks available in total — as above]

   b) P(Paul wins) = P(Paul wins 3-0) + P(Paul wins 2-1)
   $= \left(\frac{2}{3} \times \frac{2}{3} \times \frac{2}{3}\right) + \left(\frac{2}{3} \times \frac{2}{3} \times \frac{1}{3}\right) + \left(\frac{2}{3} \times \frac{1}{3} \times \frac{2}{3}\right) + \left(\frac{1}{3} \times \frac{2}{3} \times \frac{2}{3}\right)$
   $= \frac{8}{27} + \frac{4}{27} + \frac{4}{27} + \frac{4}{27} = \frac{20}{27}$
   [3 marks available — 1 mark for finding the probability of Paul winning 3-0, 1 mark for finding the probabilities of Paul winning 2-1, 1 mark for the correct final answer]
   You could draw another tree diagram showing 3 rolls if you're struggling to find the right probabilities.

3 a)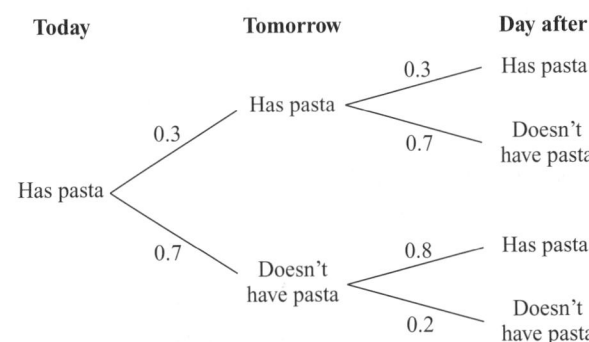

[2 marks available — 1 mark for the correct probabilities for tomorrow, 1 mark for the correct probabilities for the day after]

b) P(pasta on 1 of the next 2 days) = P(pasta then no pasta) + P(no pasta then pasta)
= (0.3 × 0.7) + (0.7 × 0.8) [1 mark]
= 0.21 + 0.56 = 0.77 [1 mark]
[2 marks available in total — as above]

4 P(different fruit) = 1 − P(same fruit)
P(same fruit) = P(two apples) + P(two oranges) + P(two pears)
= $\left(\frac{6}{20} \times \frac{5}{19}\right) + \left(\frac{9}{20} \times \frac{8}{19}\right) + \left(\frac{5}{20} \times \frac{4}{19}\right)$
[1 mark for correct probabilities, 1 mark for the correct sum]
= $\frac{122}{380} = \frac{61}{190}$ [1 mark]
P(different fruit) = 1 − $\frac{61}{190} = \frac{129}{190}$ [1 mark]
[4 marks available in total — as above]
You could also have worked out the probabilities for every choice where the fruit are different and added them together, but that would be a lot more work.

Page 106: Probability from Venn Diagrams

1 a) The number of people who only bought food
= 1320 − 1080 = 240 people
The number of people who only bought a drink
= 2560 − 1080 = 1480 people
[1 mark for either correct number]
The number of people who didn't buy either
= 4000 − 240 − 1080 − 1480 = 1200 [1 mark]

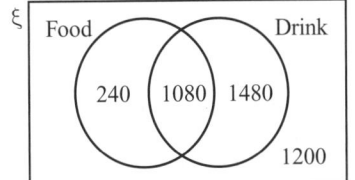

[1 mark for a correct Venn diagram]
[3 marks available in total — as above]

b) P(food but no drink) = $\frac{240}{4000} = \frac{3}{50}$ [1 mark]

2 a)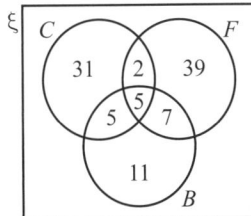

[3 marks available — 3 marks for a completely correct diagram, otherwise 1 mark for at least 2 correct entries or 2 marks for at least 4 correct entries]

b) $\frac{31 + 5 + 5 + 2 + 7 + 39}{100} = \frac{89}{100}$ or 0.89 [1 mark]

c) $\frac{7 + 5}{39 + 2 + 5 + 7}$ [1 mark] = $\frac{12}{53}$ [1 mark]
[2 marks available in total — as above]

Page 107: Relative Frequency

1 a) Relative frequency of hitting the target with a left-handed throw = $\frac{12}{20} = \frac{3}{5}$ or 0.6.
[2 marks available — 1 mark for a correct method, 1 mark for the correct answer]

b) E.g. The estimated probability is more reliable for his right hand because he threw the ball more times with that hand.
[1 mark]

2 a)
Number on counter	1	2	3	4	5
Frequency	23	25	22	21	9
Relative Frequency	0.23	0.25	0.22	0.21	0.09

[2 marks available — 2 marks for all correct answers, otherwise 1 mark for any frequency ÷ 100]

b) Elvin is likely to be wrong. The bag seems to contain fewer counters numbered 5. [1 mark]

c) P(odd number) = 0.23 + 0.22 + 0.09 [1 mark]
= 0.54 [1 mark]
[2 marks available in total — as above]

Page 108: Organising Data

1 a) Qualitative [1 mark]
b) Discrete [1 mark]

2 a) Any two from:
No time frame is given. / The response boxes do not cover all possible outcomes. / The response boxes overlap.
[2 marks available — 1 mark for each sensible answer]

b) E.g. Xin has chosen suitable inequalities because none of the classes overlap and they cover all possible options.
[2 marks available — 1 mark for identifying the classes don't overlap, 1 mark for identifying the classes cover all options]

c) E.g. Xin's results are likely to be biased because she hasn't selected teenagers at random — she's only asked her friends. So Xin can't use her results to draw conclusions about all teenagers.
[1 mark for a correct comment with suitable reasoning]
You could also say that her sample is too small to represent all teenagers.

Page 109: Mean, Median, Mode and Range

1 a) In ascending order: 3, 10, 12, 12, 13, 18, 25, 33, 37, 41
Median = (10 + 1) ÷ 2 = 5.5th value [1 mark]
= (13 + 18) ÷ 2 = 15.5 minutes [1 mark]
[2 marks available in total — as above]

b) Mean = (3 + 10 + 12 + 12 + 13 + 18 + 25 + 33 + 37 + 41) ÷ 10
= 204 ÷ 10 [1 mark] = 20.4 minutes
= 20 minutes (to the nearest minute) [1 mark]
[2 marks available in total — as above]

2 a) True — the mean number is higher than 17 because the 11th data value is higher than the mean of the original 10 values.
[1 mark]

b) False — you can't tell if the median number is higher than 15, because you don't know the other data values.
[1 mark]

3 a) 23, 26, 36 (in any order)
range = 13, median = 26
[2 marks available — 1 mark for all three weights correct, 1 mark for both range and median correct]

b) 32 + 23 + 31 + 28 + 36 + 26 = 176
4 × 27.25 = 109 [1 mark]
176 − 109 = 67 [1 mark]
so, goats weighing 31 kg and 36 kg [1 mark]
[3 marks available in total — as above]

Page 110: Frequency Tables

1. a) 5 − 0 = 5 *[1 mark]*
   No one had 6 pets so you don't include 6 in the calculation.

   b)
Number of pets	Frequency	No. of pets × Frequency
0	8	0
1	3	3
2	5	10
3	8	24
4	4	16
5	1	5
6	0	0
Total		58

   Mean = 58 ÷ 29 = 2
   [3 marks available — 1 mark for multiplying the numbers by the frequencies, 1 mark for finding the correct total number of pets, 1 mark for the correct final answer]

   c) Position of median = (29 + 1) ÷ 2 = 15 *[1 mark]*
   11th value = 1 and 12th to 16th value = 2, so median = 2 *[1 mark]*
   [2 marks available in total — as above]

2. a) 100 *[1 mark]*
   The mode is the category with the greatest frequency.

   b) Position of median = (180 + 1) ÷ 2 = 90.5 *[1 mark]*
   There are 6 + 20 + 44 = 70 bags in the first three columns and 70 + 108 = 178 bags in the first four columns.
   So the 90.5 position is in the fourth column.
   So the median number of nails per bag is 100 *[1 mark]*.
   [2 marks available in total — as above]

   c) (97 × 6) + (98 × 20) + (99 × 44)
           + (100 × 108) + (101 × 2) = 17 900
   17 900 ÷ 180 = 99.44... = 99.4 (1 d.p.)
   [3 marks available — 1 mark for multiplying the numbers by the frequencies, 1 mark for finding the correct total number of nails, 1 mark for the correct final answer]

Page 111: Grouped Frequency Tables

1. a) The highest frequency is 8, so the modal class is $40 < x \leq 50$ *[1 mark]*

   b) 70 − 10 = 60 *[1 mark]*
   This is the largest possible range (assuming someone got a mark of 10 and someone else got a mark 70). In reality, the range is likely to be smaller.

   c)
Exam mark, x	Freq.	Mid-interval value	Frequency × mid-interval value
$10 < x \leq 20$	2	(10 + 20) ÷ 2 = 15	2 × 15 = 30
$20 < x \leq 30$	4	(20 + 30) ÷ 2 = 25	4 × 25 = 100
$30 < x \leq 40$	7	(30 + 40) ÷ 2 = 35	7 × 35 = 245
$40 < x \leq 50$	8	(40 + 50) ÷ 2 = 45	8 × 45 = 360
$50 < x \leq 60$	3	(50 + 60) ÷ 2 = 55	3 × 55 = 165
$60 < x \leq 70$	6	(60 + 70) ÷ 2 = 65	6 × 65 = 390

   Mean = (30 + 100 + 245 + 360 + 165 + 390) ÷ 30
   = 1290 ÷ 30 = 43
   [4 marks available — 1 mark for all mid-interval values, 1 mark for calculation of frequency × mid-interval value, 1 mark for dividing sum of frequency × mid-interval values by sum of frequencies, 1 mark for the correct final answer]

2. a) ((24 × 4) + (28 × 8) + (32 × 13) + (36 × 6) + (40 × 1)) ÷ 32
   = 992 ÷ 32 = 31 seconds
   [4 marks available — 1 mark for all mid-interval values, 1 mark for calculation of frequency × mid-interval value, 1 mark for dividing sum of frequency × mid-interval values by sum of frequencies, 1 mark for the correct answer]

   b) There were 32 pupils and 13 + 6 + 1 = 20 got a time of more than 30 seconds *[1 mark]*, (20 ÷ 32) × 100 = 62.5% *[1 mark]*
   [2 marks available in total — as above]

   c) E.g. You couldn't use these results because you don't know the ages of the pupils in the sample, or whether any of the times were run by boys, so you can't tell if the results would fairly represent 16-year-old boys at the school.
   [1 mark for a sensible comment]

Pages 112-113: Simple Charts

1. a) 55 − 20 = 35 *[1 mark]*

   b) Saturday = 25 + 20 + 35 + 20 = 100 cups
   Sunday = 40 + 55 + 10 + 5 = 110 cups
   So more hot drinks were sold on Sunday.
   [2 marks available — 1 mark for finding the correct total for either day, 1 mark for the correct answer]

   c) 25 cups of herbal tea were sold in total, and $\frac{20}{25} = \frac{4}{5}$ of them were sold on Saturday *[1 mark]*.

2. a) 40 − 20 = 20 *[1 mark]*

   b)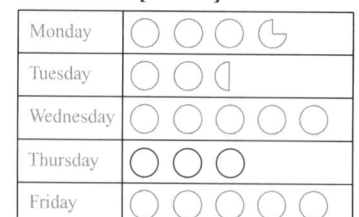
   [1 mark]

   c) Number of eggs laid on Friday = 5 × 8 = 40 *[1 mark]*
   0.4 × 40 = 16 eggs *[1 mark]*
   [2 marks available in total — as above]

3. a) 3 days *[1 mark]*

   b) They sold more than 30 newspapers on 3 days and there are 30 days in total. So the fraction is $\frac{3}{30} = \frac{1}{10}$. *[1 mark]*

   c) There are 30 data values so the median is the (30 + 1) ÷ 2 = 15.5th value. *[1 mark]*
   The 15th value is 15 and the 16th value is 17 so the median is (17 + 15) ÷ 2 = 16. *[1 mark]*
   [2 marks available in total — as above]

4. There are 4 + 2.5 + 3.5 = 10 symbols in total *[1 mark]*
   So 1 symbol represents 100 ÷ 10 = 10 jars of jam *[1 mark]*
   3.5 × 10 *[1 mark]* = 35 jars of raspberry jam *[1 mark]*
   [4 marks available in total — as above]

5. a) Total = (4 × 0) + (10 × 1) + (8 × 2) + (2 × 3) + (1 × 4) *[1 mark]*
   = 36
   Mean = 36 ÷ 25 *[1 mark]* = 1.44 *[1 mark]*
   [3 marks available in total — as above]

   b) Median = (25 + 1) ÷ 2 = 13th value *[1 mark]*.
   13th value is shown by the 2nd bar, so median = 1 *[1 mark]*.
   [2 marks available in total — as above]

Page 114: Pie Charts

1. a) Total number of people = 12 + 18 + 9 + 21 = 60
   Multiplier = 360 ÷ 60 = 6
   Plain: 12 × 6 = 72°
   Salted: 18 × 6 = 108°
   Sugared: 9 × 6 = 54°
   Toffee: 21 × 6 = 126°

   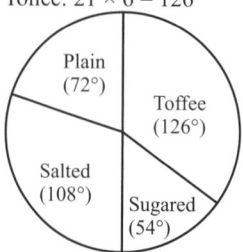

   [4 marks available — 1 mark for one sector correctly drawn, 1 mark for a second sector correctly drawn, 1 mark for a complete pie chart with all angles correct, 1 mark for correct labels]

   b) E.g. Chris is not right because there is no information about the number of people in the ice-cream survey. *[1 mark]*

2   There are 360° in a circle, so
    $2x + 3x + 4x + x + 90° = 360°$
    $10x = 270°$, so $x = 27°$
    The sector for leek & potato is $3x = 3 \times 27 = 81°$,
    so $\frac{81}{360} \times 80 = 18$ students chose leek & potato soup.
    [4 marks available — 1 mark for forming an equation in terms of x, 1 mark for solving to find the value of x, 1 mark for a correct method to find the number of students, 1 mark for the correct answer]

Page 115: Scatter Diagrams

1   a)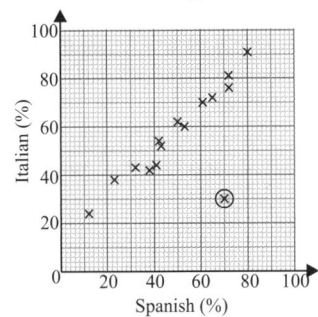
    [1 mark]
    b) Strong positive correlation *[1 mark]*
    c)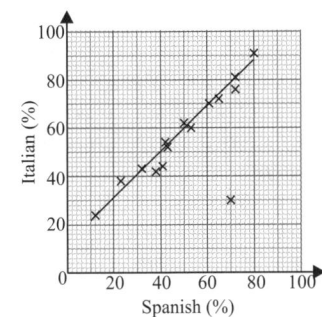
    [1 mark for line of best fit passing between (10, 16) & (10, 28) and (80, 82) & (80, 96)]

2   a)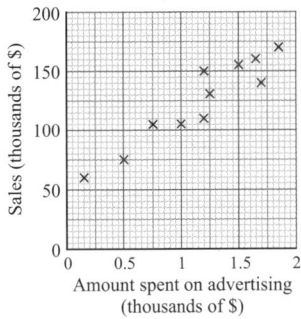
    [1 mark if all three points are plotted correctly]
    b) As the amount spent on advertising increases, so does the value of sales. *[1 mark]*
    Or you could say there's a positive correlation between the amount spent and the value of sales.
    c)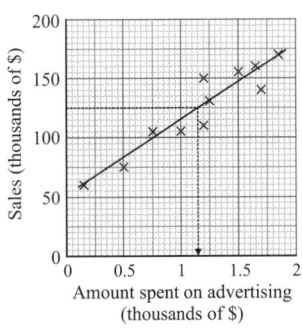
    See graph — $1150
    [2 marks available — 1 mark for drawing a line of best fit, 1 mark for reading off the correct answer, allow answers ± $100]

Pages 116-117: Interpreting and Comparing Data

1   a) Correct as only two of Group A had no favourite flavour whereas 15 of Group B did. *[1 mark]*
    b) Incorrect because, e.g., salt & vinegar are more popular than cheese & onion in both groups / a similar proportion of each group preferred ready salted / a similar proportion of each group preferred other flavours / other flavours were more popular in each group than any of the specified ones. *[1 mark]*
    Any similarity you can find between the two groups will earn you a mark, as it contradicts the statement in the question.

2   E.g. The horizontal axis is angled downwards, which exaggerates the increasing trend of the line *[1 mark]*. The vertical axis is broken, which makes the rise in the line seem steeper *[1 mark]*.
    [2 marks available in total — as above]

3   The median *[1 mark]*, since there is an outlier in the data *[1 mark for a sensible reason]*.
    [2 marks available in total — as above]
    The outlier would have a big impact on the mean because it is so far from the other values and the data set is small. There are also multiple modes which, in this case, makes it less useful than the median.

4   a) E.g. the runs scored by cricketer A are fairly evenly spread, so they were approximately equally likely to score any number of runs in the range.
    Cricketer B tends to get more runs above 30 than below, so they scored a higher number of runs more often.
    [2 marks available — 1 mark for each of two valid points]
    b) Cricketer B — e.g. more of cricketer B's data is towards the higher end of the diagram (in the 30s and 40s).
    [1 mark for any sensible reason]
    You could also compare the medians of the two data sets.

5   The interquartile range *[1 mark]*, because the data value of 4 is an outlier and so would have a big effect on the range *[1 mark for a sensible explanation]*.
    [2 marks available in total — as above]

6   a) Rewrite the data in ascending order:
    3, 4, 4, 5, 5, 6, 7, 7, 8, 8, 9, 10, 11, 11, 12
    The lower quartile is the $(15 + 1) \div 4 = 4$th value
    So the lower quartile is 5. *[1 mark]*
    The upper quartile is the $3(15 + 1) \div 4 = 12$th value
    So the upper quartile is 10. *[1 mark]*
    [2 marks available in total — as above]
    b) E.g. The interquartile range will remain the same, as all the values have decreased by 50p. This 50p will cancel out when you subtract the lower quartile from the upper quartile. *[1 mark]*.

Pages 118-119: Histograms

1   To find the scale, compare the height of the first bar to the frequency: 7.5 squares = 15 pupils, so 1 square = 2 pupils.

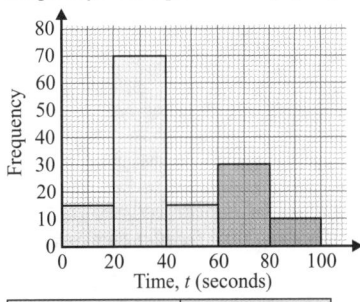

Time, t (s)	Frequency
$0 < t \leq 20$	15
$20 < t \leq 40$	70
$40 < t \leq 60$	15
$60 < t \leq 80$	30
$80 < t \leq 100$	10

[4 marks available — 1 mark for the correct scale on the frequency axis, 1 mark for the correct entry in the table, 1 mark for each correct bar on the histogram]

2   This question is asking you to estimate the mean amount of time the children watched TV for, then compare that to the mean time for the adults.

Time, m (minutes)	Frequency (f)	Mid-interval Value (x)	fx
$40 \leq m < 60$	$20 \times 1 = 20$	50	1000
$60 \leq m < 70$	$10 \times 7 = 70$	65	4550
$70 \leq m < 80$	$10 \times 4 = 40$	75	3000
$80 \leq m < 120$	$40 \times 2 = 80$	100	8000
$120 \leq m < 140$	$20 \times 3 = 60$	130	7800
Total	270		24 350

Mean for children = 24 350 ÷ 270 = 90.185...
= 90.2 minutes (to 1 d.p.)

E.g. the data supports the hypothesis since the mean time for the adults is longer than the mean time for the children, and the large samples mean the results should represent the population.
[4 marks available — 1 mark for a correct method to find the frequencies, 1 mark for multiplying the frequencies by the mid-interval values, 1 mark for the correct mean, 1 mark for a correct conclusion based on a comparison of the means]

3 a) Estimate of number of lambs between 3.5 and 4 kg
       $= 0.5 \times 22$ *[1 mark]* $= 11$
       $11 + (1 \times 26) + (1 \times 16) + (2 \times 3)$ *[1 mark]*
       $= 59$ out of $100 = 59\%$ *[1 mark]*
       [3 marks available in total — as above]

   b)

Weight, w kg	$0 < w \leq 2$	$2 < w \leq 4$	$4 < w \leq 5$	$5 < w \leq 6$	$6 < w \leq 8$
Frequency	4	28	30	28	10
Frequency Density	2	14	30	28	5

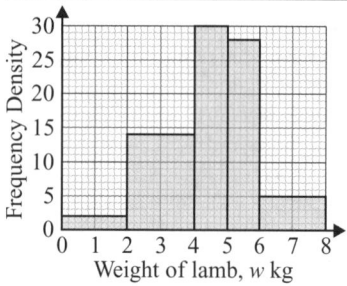

[4 marks available — 1 mark for all bars with correct widths. 3 marks for all bars with correct heights, otherwise 2 marks for at least two bars with correct heights, or 1 mark for one bar with the correct height]

   c) E.g. the second histogram shows more lambs with heavier weights and fewer with lighter weights than the first, which suggests there is a difference between the two farms.
       [1 mark for a correct comment based on a comparison of the histograms]

Pages 120-121: Cumulative Frequency

1 a)

Time taken (t minutes)	$1.5 \leq t < 2$	$2 \leq t < 2.5$	$2.5 \leq t < 3$	$3 \leq t < 3.5$	$3.5 \leq t < 4$
Frequency	10	36	34	14	6
Cumulative Frequency	10	46	80	94	100

[2 marks available — 1 mark for the correct missing frequency, 1 mark for all cumulative frequencies correct]

   b)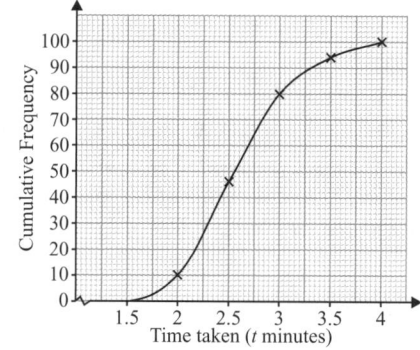

[2 marks available — 1 mark for plotting the points correctly, 1 mark for connecting the points with a curve or straight lines]

   c) The median time is when the cumulative frequency is 100 ÷ 2 = 50 *[1 mark]*. So read across from 50 and down to get 2.55 minutes (allow 2.6) *[1 mark]*.
       [2 marks available in total — as above]

   d) The cumulative frequency at 2 minutes is 10, and the cumulative frequency at 3.4 minutes is 92. So 92 − 10 = 82 kettles took between 2 and 3.4 minutes, which is $\frac{82}{100} = \frac{41}{50}$.
       [3 marks available in total — 1 mark for each correct value read from your graph, 1 mark for the correct answer]

2 a) (i) The lower quartile is at 80 ÷ 4 = 20, which is 4.8 m.
            The upper quartile is at (80 ÷ 4) × 3 = 60, which is 6.6 m.
            So interquartile range = 6.6 − 4.8 = 1.8 m.
            [2 marks available — 2 marks for the correct answer, otherwise 1 mark for finding the correct value of the lower or upper quartile]

       (ii) The 65th percentile is at 80 × 0.65 = 52, *[1 mark]* which is 6.2 m *[1 mark]*
            [2 marks available in total — as above]

       (iii) Read up from 8 m and across to a cumulative frequency of 74 *[1 mark]*, so there are 80 − 74 = 6 trees taller than 8 m. *[1 mark]*
            [2 marks available in total — as above]

   b) Approximately 11 trees are shorter than 4 m, and 48 trees are shorter than 6 m, so 48 − 11 = 37 trees are between 4 m and 6 m. *[1 mark]* As a percentage, this is (37 ÷ 80) × 100 = 46.25% of the trees, so yes, the diagram supports Saul's claim. *[1 mark]*
       [2 marks available in total — as above]

Pages 122-135 — Mixed Questions

1 When Natalie squares her number the final digit is 1, so her number must end in 1 or 9. *[1 mark]*
   It must be either 11, 19, 21 or 29, but her number is not prime so it must be 21. *[1 mark]*
   [2 marks available in total — as above]

2 a) Elements of A are 3, 4, 5, 6
       Elements of B are 1, 2, 3, 4, 6 (12 is not in the universal set ξ)

       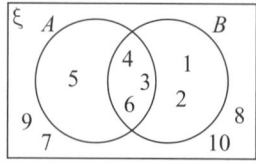

       [2 marks available — 2 marks for a completely correct Venn diagram, otherwise 1 mark if a maximum of 2 elements are in the wrong set]

   b) $A \cup B = \{1, 2, 3, 4, 5, 6\}$, so n($A \cup B$) = 6 *[1 mark]*

   c) There are 4 elements in set A and 3 are also in set B:
       So P(in B given in A) = $\frac{3}{4}$ *[1 mark]*

3 $3\frac{3}{5} \div \frac{4}{7} = \frac{18}{5} \div \frac{4}{7} = \frac{18}{5} \times \frac{7}{4} = \frac{9}{5} \times \frac{7}{2} = \frac{63}{10} = 6\frac{3}{10}$
   [3 marks available — 1 mark for taking the reciprocal and multiplying the two fractions together, 1 mark for an equivalent fraction, 1 mark for the correct final answer]

4 a) 4th term = $4^2 + (2 \times 4) + 5 = 16 + 8 + 5 = 29$ so Fran is correct
       [2 marks available — 2 marks for indicating that Fran is correct and finding that the 4th term is 29, otherwise 1 mark for correct method to find the 4th term]
   b) A term in the sequence is the sum of the two previous terms. The next two terms are $12 + 19 = 31$ *[1 mark]* and $19 + 31 = 50$ *[1 mark]*
       [2 marks available in total — as above]

5 a) $AE = 12$ and $BD = 18$
       $AE : BD = 12 : 18 = 2 : 3$
       [2 marks available — 1 mark for finding the lengths of the two sides, 1 mark for the correct answer]
   b) $DC = 27 - 12 = 15$ *[1 mark]*
       From a) you know that corresponding sides of the two similar triangles are in the ratio 2 : 3.
       $2 : 3 = BE : CD = BE : 15$ so $BE = 10$ *[1 mark]*
       The coordinates of E are $(12, 18 - 10) = (12, 8)$. *[1 mark]*
       [3 marks available in total — as above]

6 a) Expected number = $(3.1 \times 10^{-4}) \times (7 \times 10^5)$ *[1 mark]*
       $= 217 = 2.17 \times 10^2$ screws *[1 mark]*
       [2 marks available in total — as above]
   b) Let $r = 0.0\dot{4}$, so $10r = 0.\dot{4}$
       $100r = 4.\dot{4}$ *[1 mark]*
       $90r = 4.\dot{4} - 0.\dot{4} = 4$
       $r = \frac{4}{90} = \frac{2}{45}$ *[1 mark]*
       [2 marks available in total — as above]

7 a) $2a + b = 10 \xrightarrow{\times 2} 4a + 2b = 20$ *[1 mark]*
       $4a + 2b = 20$ $2(3) + b = 10$
       $\underline{3a + 2b = 17\ -}$ $b = 10 - 6$
       $a = 3$ *[1 mark]* $b = 4$ *[1 mark]*
       [3 marks available in total — as above]
   b) $a\begin{pmatrix}2\\1\end{pmatrix} - b\begin{pmatrix}3\\2\end{pmatrix} = 3\begin{pmatrix}2\\1\end{pmatrix} - 4\begin{pmatrix}3\\2\end{pmatrix} = \begin{pmatrix}6\\3\end{pmatrix} - \begin{pmatrix}12\\8\end{pmatrix}$ *[1 mark]*
       $= \begin{pmatrix}-6\\-5\end{pmatrix}$ *[1 mark]*
       [2 marks available in total — as above]

8 a) Angle for car = $360° - 162° - 36° - 45° = 117°$ *[1 mark]*
       Number of office assistants that travel by car is $= \frac{117}{360} \times 80$
       $= 26$ *[1 mark]*
       [2 marks available in total — as above]
   b) Total number of staff = $80 + 10 = 90$
       Number that travel by car = $26 + 10 = 36$
       So $\frac{36}{90} \times 100 = 40\%$ travel by car.
       [2 marks available — 2 marks for the correct answer, otherwise 1 mark for finding both the total number of staff and the number that travel by car]
   c) Dave's speed is $28 \div (42 \div 60) = 40$ km/h *[1 mark]*
       Olivia's speed is $40 + 10 = 50$ km/h
       So it takes her $28 \div 50 = 0.56$ hours *[1 mark]*
       $0.56 \times 60 = 33.6 = 34$ minutes (to the nearest minute) *[1 mark]*
       [3 marks available in total — as above]
   d) E.g. If Olivia drives a different route then she might go a different distance, which could make her journey time longer or shorter. *[1 mark]*

9 a) (i) 1 cupcake needs $132 \div 12 = 11$ g sugar.
            So 30 cupcakes need $30 \times 11 = 330$ g sugar.
            [2 marks available — 1 mark for a correct method, 1 mark for the correct answer]
       (ii) Upper bound = $350.0 + 0.05 = 350.05$ g
            Lower bound = $350.0 - 0.05 = 349.95$ g
            So $349.95 \leq f < 350.05$
            [2 marks available — 1 mark for each correct bound]
   b) Multiples of 12: 12, 24, 36, 48, 60 …
       Multiples of 10: 10, 20, 30, 40, 50, 60, …
       The smallest number of each type of biscuit Ajay could have bought is 60. This is $60 \div 10 = 6$ packets of shortbread biscuits.
       [2 marks available — 1 mark for the correct method to the find the lowest common multiple, 1 mark for correct answer]
   c) Multiplier = $1 + 0.004 = 1.004$ *[1 mark]*
       2 years = $2 \times 12 = 24$ months
       $\$1.79 \times (1.004)^{24} = \$1.9699…$
       $= \$1.97$ (to the nearest cent) *[1 mark]*
       [2 marks available in total — as above]

10 a) Total number of cars = 12
       Number of cars with a maximum speed less than 180 km/h = 5 *[1 mark]*
       Percentage = $(5 \div 12) \times 100 = 41.67\%$ (2 d.p.) *[1 mark]*
       [2 marks available in total — as above]
   b) Strong positive correlation *[1 mark]*
   c) Ignore the outlier when drawing a line of best fit.

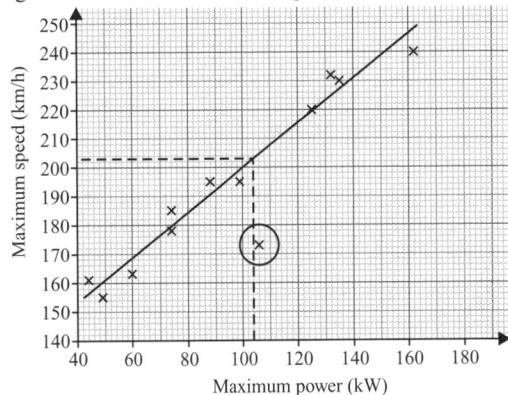

       Maximum speed = 203 km/h (allow ± 5).
       [2 marks available — 1 mark for drawing a line of best fit (ignoring the outlier), 1 mark for accurately reading from your graph the speed corresponding to a power of 104 kW]
   d) Choose two points on your line and find the gradient between them, e.g. for the points (162, 248) and (100, 200):
       Gradient $= \frac{(248 - 200)}{(162 - 100)}$ *[1 mark]*
       $= 0.774… = 0.8$ (1 d.p.) *[1 mark]*
       (allow 0.7-0.9 based on line of best fit that is drawn)
       [2 marks available in total — as above]
   e) 190 kW lies outside of the range of data plotted on the scatter graph. *[1 mark]*

11 a) Exterior angle of a regular nine-sided shape
       $= \frac{360°}{9} = 40°$ *[1 mark]*
       So angle $ABC = 2 \times 40° = 80°$ *[1 mark]*
       Angle $ADC = 360° - 90° - 90° - 80° = 100°$ *[1 mark]*
       [3 marks available in total — as above]
       Alternatively you could find the size of each interior angle (140°) of the regular nine-sided shape and use this to find angle ABC.
   b) (i) $2 + 3 + 5 = 10$ parts in the ratio.
            Angles in a triangle add to 180°, so 10 parts = 180° and 1 part = 180° ÷ 10 = 18°. *[1 mark]*
            So angle $EFG = (2 \times 18°) = 36°$,
            angle $EGF = (3 \times 18°) = 54°$ and
            angle $FEG = (5 \times 18°) = 90°$. *[1 mark]*
            [2 marks available in total — as above]
       (ii) $\sin x = \frac{O}{H}$, so $\sin 36° = \frac{EG}{15}$ *[1 mark]*
            $EG = 15 \times \sin 36° = 8.8167…$ cm *[1 mark]*
            EG is a side of the regular polygons, and $ABCD$ is a symmetrical shape, so $AD = DC$.
            Total perimeter = $(14 \times 8.8167…) + (2 \times 7.4)$ *[1 mark]*
            $= 138.2349… = 138.2$ cm *[1 mark]*
            [4 marks available in total — as above]

12 Rationalise the denominator of the first term:
$\frac{6}{\sqrt{3}} = \frac{6\sqrt{3}}{\sqrt{3} \times \sqrt{3}} = \frac{6\sqrt{3}}{3} = 2\sqrt{3}$ *[1 mark]*

And simplify the second term:
$\sqrt{27} = \sqrt{9}\sqrt{3} = 3\sqrt{3}$ *[1 mark]*
So $\frac{6}{\sqrt{3}} + \sqrt{27} = 2\sqrt{3} + 3\sqrt{3} = 5\sqrt{3}$ *[1 mark]*
[3 marks available in total — as above]

13 a) $256 = 4 \times 4 \times 4 \times 4 = 4^4$
So $x = 4$ *[1 mark]*

b) $81 = 9 \times 9 = 3 \times 3 \times 3 \times 3 = 3^4$ *[1 mark]*
So $81^{\frac{1}{4}} = 3$
$x = \frac{1}{4}$ *[1 mark]*
[2 marks available in total — as above]

14 The upper bound for the radius is $16 + 0.5 = 16.5$ cm.
The upper bound for the height is $20 + 0.5 = 20.5$ cm.
[1 mark for both upper bounds]
Maximum volume of cone $= \frac{1}{3}\pi \times 16.5^2 \times 20.5$
$= 5844.5404...$ cm^3 *[1 mark]*
The lower bound for the rate that water leaves is:
$0.39 - 0.005 = 0.385$ litres per minute
$= 385$ cm^3 per minute *[1 mark]*
So upper bound for time $= 5844.5404... \div 385$
$= 15.1806...$ minutes *[1 mark]*
So Marion is not correct because it could take up to 15.1806... minutes to empty. *[1 mark]*
[5 marks available in total — as above]

15 Let Hannah's number be n, then Tim's number is $n + 1$
Form the equation: $\frac{1}{n} + \frac{1}{n+1} = \frac{5}{6}$ *[1 mark]*
Therefore, $\frac{n+1}{n(n+1)} + \frac{n}{n(n+1)} = \frac{5}{6}$
$\frac{2n+1}{n(n+1)} = \frac{5}{6}$
$6(2n + 1) = 5n(n + 1)$ *[1 mark]*
$12n + 6 = 5n^2 + 5n$
$5n^2 - 7n - 6 = 0$ *[1 mark]*
$(5n + 3)(n - 2) = 0$ *[1 mark]*
So $n = -\frac{3}{5}$ or $n = 2$
Hannah's number is negative so she was thinking of $-\frac{3}{5}$. *[1 mark]*
[5 marks available in total — as above]
The quadratic formula could be used to solve the equation.

16 a) $y = Ax^n$ is a positive square root graph, so $n = \frac{1}{2}$. *[1 mark]*

b) $y = Ax^{-1}$ *[1 mark]*
Reciprocal graphs of the form $y = Ax^{-1}$ have two lines of symmetry.

17 a) Change in $y = 8 - (-7) = 15$
Change in $x = 3 - (-2) = 5$
So gradient $= 15 \div 5 = 3$
So $y = mx + c$ becomes $y = 3x + c$
Put in $x = 3$ and $y = 8$ to find the value of c:
$8 = 3(3) + c$, which means $c = 8 - 9 = -1$
The equation of the line is $y = 3x - 1$.
[3 marks available — 1 mark for the correct gradient, 1 mark for putting one point into the equation, 1 mark for the correct answer]

b) $\left(\frac{-2+3}{2}, \frac{-7+8}{2}\right) = \left(\frac{1}{2}, \frac{1}{2}\right)$
[2 marks available — 1 mark for a correct method, 1 mark for the correct answer]

c)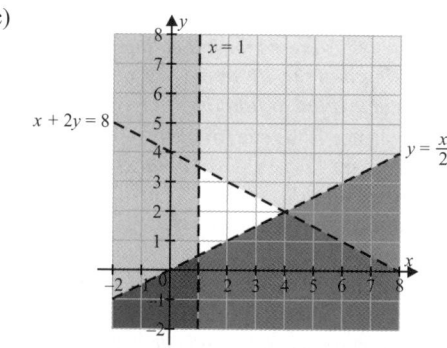
[4 marks available — 1 mark for drawing each line correctly, 1 mark for shading the unwanted regions correctly]

18 $\frac{10}{11} = \frac{90}{99}$ *[1 mark]* $= 0.\dot{9}\dot{0}$ *[1 mark]*
[2 marks available in total — as above]

19 a) Angle ABC must be a right angle because side AC of the triangle is a diameter (angle in a semicircle).
$\cos 52° = \frac{14.2}{AC}$ *[1 mark]*
So $AC = \frac{14.2}{\cos 52°} = 23.0646...$ cm *[1 mark]*
Circumference $= \pi \times 23.0646...$
$= 72.4596...$ cm $= 72.5$ cm (3 s.f.) *[1 mark]*
[3 marks available in total — as above]

b) Angle GED $= 35°$ *[1 mark]* (alternate segment theorem)
Angle ODG $= 90° - 35° = 55°$ *[1 mark]*
(a tangent and a radius are perpendicular)
Obtuse angle FOD $= 2 \times$ angle FED
$= 2(39° + 35°) = 2 \times 74° = 148°$ *[1 mark]*
(angle at centre is twice angle at circumference)
Angle FGD $= 180° - 74° = 106°$ *[1 mark]*
(opposite angles in a cyclic quadrilateral add up to 180°)
Angle GFO $= 360° - 106° - 148° - 55° = 51°$ *[1 mark]*
(angles in a quadrilateral add up to 360°)
[5 marks available in total — as above]
There may be alternative ways of getting to the same answer.

20 a)
x	−4	−3	...	0.5	...
f(x)	−9.4375	−7	...	35	...

[2 marks available — 1 mark for each correct value]

b)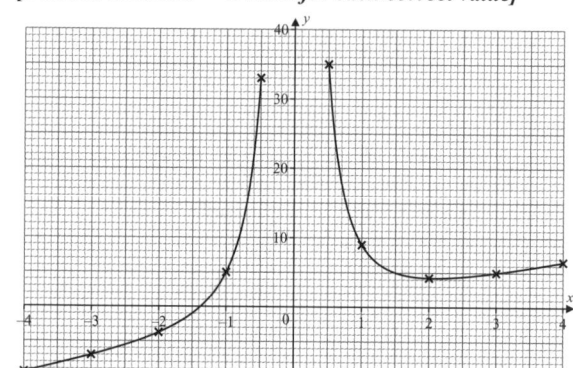

[5 marks available — 5 marks for a correct smooth curve, otherwise 4 marks for the correct curve with the branches joined, 3 marks for 9 or 10 points plotted correctly, 2 marks for 7 or 8 points plotted correctly, 1 mark for 5 or 6 points plotted correctly]

c) Draw a tangent to the curve at $x = 1.5$ and find the gradient.

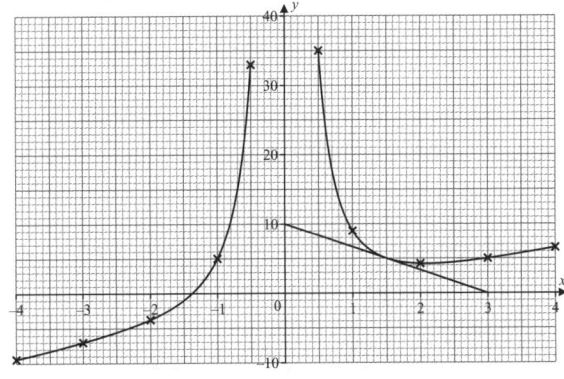

Gradient = $\frac{10 - 0}{0 - 3}$ = –3.3333... = –3.3 (1 d.p.)

[2 marks available — 1 mark for the correct method, 1 mark for a gradient between –3 and –4]

d) $\frac{9}{x^2} + x - 10 = 0 \Rightarrow \frac{9}{x^2} + 2x - 2 = x + 8$

So draw the line $y = x + 8$ on the graph.

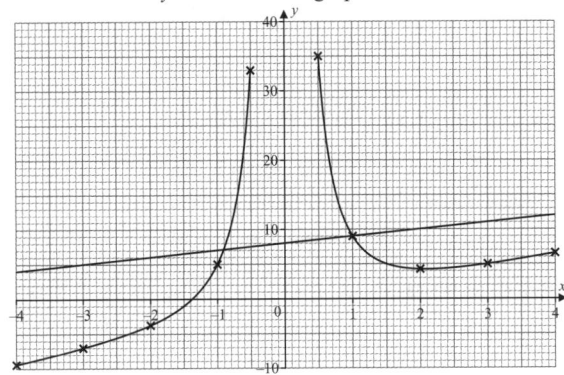

The solution is where $y = x + 8$ intersects the graph, so $x = -0.9$ (allow –1.0 or –0.8) or 1 (allow 0.9 or 1.1).

[3 marks available — 1 mark for finding the correct equation for the line, 1 mark for plotting the line correctly, 1 mark for both correct values of x]

21 a) Calculate frequency densities using frequency ÷ class width

Age (A years)	Frequency	Frequency Density
$18 \leq A < 20$	18	$18 \div 2 = 9$
$20 \leq A < 25$	35	$35 \div 5 = 7$
$25 \leq A < 30$	40	$40 \div 5 = 8$
$30 \leq A < 40$	45	$45 \div 10 = 4.5$
$40 \leq A < 60$	50	$50 \div 20 = 2.5$
$60 \leq A < 70$	75	$75 \div 10 = 7.5$
$70 \leq A < 90$	40	$40 \div 20 = 2$

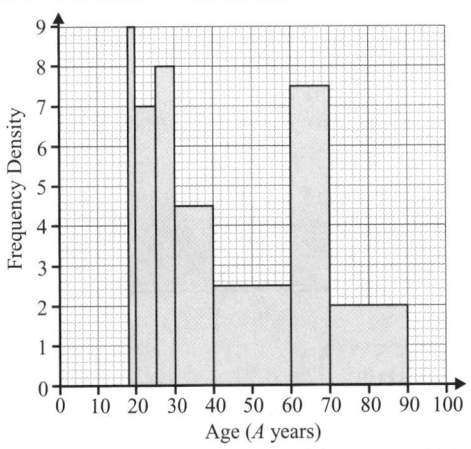

[5 marks available — 5 marks for a correct histogram, otherwise award 1 mark for a correct method to find frequency densities, 1 mark for all the correct frequency densities, 1 mark for y-axis labelled appropriately, 1 mark for at least 3 bars drawn correctly]

b)

Age (A years)	Frequency	Mid-interval values	Mid-interval value × F
$18 \leq A < 20$	18	19	342
$20 \leq A < 25$	35	22.5	787.5
$25 \leq A < 30$	40	27.5	1100
$30 \leq A < 40$	45	35	1575
$40 \leq A < 60$	50	50	2500
$60 \leq A < 70$	75	65	4875
$70 \leq A < 90$	40	80	3200

Total frequency = 303
Total (Mid-interval value × F) = 14 379.5
Mean = 14 379.5 ÷ 303 = 47.4570...
= 47 (to the nearest whole number)

[4 marks available — 1 mark for finding the mid-interval values, 1 mark for multiplying the frequency and the mid-interval values together, 1 mark for finding the totals correctly, 1 mark for the correct answer]

c) The ages of the members have a large range. There are a lot of people with ages much lower than the mean and a lot of people with ages much higher than the mean, so the mean is not a very typical age of the members.

[1 mark for a comment like these or that refers to the mean age being in a group which has a low frequency density]

d) P(at least one over 30) = 1 – P(both under 30)
Adult gym members = 303
Adult gym members under 30 = 18 + 35 + 40 = 93
P(at least one over 30) = $1 - \left(\frac{93}{303} \times \frac{92}{302}\right)$
= 0.9064... = 0.91 (2 d.p.)

[3 marks available — 1 mark for any correct method, 1 mark for using conditional probabilities, 1 mark for the correct answer]

22 a) f(–6) = 2 × (–6) – 15 = –27 *[1 mark]*

b) $2a - 15 = 5$, so $2a = 20$, $a = 10$ *[1 mark]*

c) fg(x) = $2(x^2 + c) - 15 = 2x^2 + 2c - 15$ *[1 mark]*
fg(4) = 25, so $2 \times 4^2 + 2c - 15 = 25$
$2c = 8$, $c = 4$ *[1 mark]*
[2 marks available in total — as above]

d) h(–2) = 2 × (–2)² = 8, h(0) = 2 × 0² = 0, h(2) = 2 × 2² = 8
So the range of h(x) is: {0, 8} *[1 mark]*
You still get the mark here if you don't use set notation, so long as your answer is clear.

e) $\frac{dy}{dx} = x^2 + 4x - 12$

[2 marks available — 2 marks for the correct answer, otherwise 1 mark for differentiating two terms correctly]

f) Turning points are when $\frac{dy}{dx} = 0$.
$x^2 + 4x - 12 = 0$ *[1 mark]*
$(x - 2)(x + 6) = 0$ *[1 mark]*
So $x = 2$ and $x = -6$ *[1 mark]*

When $x = 2$, $y = (2^3 \div 3) + 2 \times 2^2 - 12 \times 2 = -\frac{40}{3}$
When $x = -6$, $y = ((-6)^3 \div 3) + 2 \times (-6)^2 - 12 \times (-6) = 72$
So the turning points are at $(2, -\frac{40}{3})$ and $(-6, 72)$. *[1 mark]*
[4 marks available in total — as above]

Formulas in the Exams

International GCSE Maths uses a lot of formulas — and you'll have a difficult time trying to answer a question without the proper formula to start you off. Fortunately, those lovely examiners give you some of the formulas you need to use:

Area of circle = πr^2
Circumference of circle = $2\pi r$

Volume of cone = $\frac{1}{3}\pi r^2 h$
Curved surface area of cone = $\pi r l$

Area of triangle = $\frac{1}{2} \times$ base $\times$ height

Volume of sphere = $\frac{4}{3}\pi r^3$
Surface area of sphere = $4\pi r^2$

Volume of prism = cross-sectional area $\times$ length

Volume of cylinder = $\pi r^2 h$
Curved surface area of cylinder = $2\pi r h$

Volume of pyramid = $\frac{1}{3} \times$ base area $\times$ height

Extended:

For any triangle ABC:

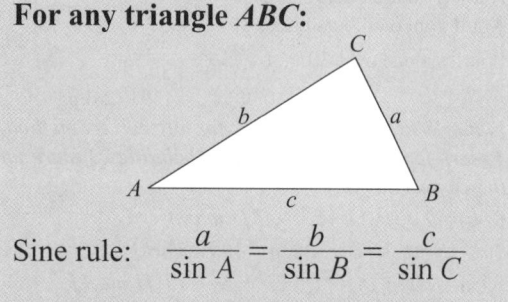

Sine rule: $\dfrac{a}{\sin A} = \dfrac{b}{\sin B} = \dfrac{c}{\sin C}$

Cosine rule: $a^2 = b^2 + c^2 - 2bc\cos A$

Area of triangle = $\frac{1}{2}ab \sin C$

The Quadratic Formula:
The solutions of $ax^2 + bx + c = 0$, where $a \neq 0$
$$x = \frac{-b \pm \sqrt{(b^2 - 4ac)}}{2a}$$

Sadly, there are lots of formulas that you're expected to be able to remember for the exam. Basically, any formulas that aren't given above, you need to learn.
There isn't space to write them all out below, but here are some of the main ones:

For a right-angled triangle:
Pythagoras' theorem: $a^2 + b^2 = c^2$
Trigonometry ratios:
$\sin x = \dfrac{O}{H}$, $\cos x = \dfrac{A}{H}$, $\tan x = \dfrac{O}{A}$

Where P(A) and P(B) are the probabilities of events A and B respectively:

P(A or B) = P(A) + P(B) (If A and B can't both happen at the same time.)

P(A and B) = P(A) × P(B) (If the result of one event doesn't affect the probability of the other.)

Compound Growth and Decay:
$N = N_0(\text{multiplier})^n$

Compound Measures:
Speed = $\dfrac{\text{Distance}}{\text{Time}}$

Area of trapezium = $\frac{1}{2}(a + b)h_v$